WHY IS THERE PHILOSOPHY
OF MATHEMATICS AT ALL?

This truly philosophical book takes us back to fundamentals – the sheer experience of proof, and the enigmatic relation of mathematics to nature. It asks unexpected questions, such as 'What makes mathematics mathematics?', 'Where did proof come from and how did it evolve?', and 'How did the distinction between pure and applied mathematics come into being?' In a wide-ranging discussion that is both immersed in the past and unusually attuned to the competing philosophical ideas of contemporary mathematicians, it shows that proof and other forms of mathematical exploration continue to be living, evolving practices – responsive to new technologies, yet embedded in permanent (and astonishing) facts about human beings. It distinguishes several distinct types of application of mathematics, and shows how each leads to a different philosophical conundrum. Here is a remarkable body of new philosophical thinking about proofs, applications, and other mathematical activities.

IAN HACKING is a retired professor of the Collège de France, Chair of Philosophy and History of Scientific Concepts, and retired University Professor of Philosophy at the University of Toronto. His most recent books include *The Taming of Chance* (1990), *Rewriting the Soul* (1995), *The Social Construction of What?* (1999), *An Introduction to Probability and Inductive Logic* (2001), *Mad Travelers* (2002), and *The Emergence of Probability* (2006).

WHY IS THERE PHILOSOPHY OF MATHEMATICS AT ALL?

IAN HACKING

CAMBRIDGE
UNIVERSITY PRESS

University Printing House, Cambridge CB2 8BS, United Kingdom

One Liberty Plaza, 20th Floor, New York, NY 10006, USA

477 Williamstown Road, Port Melbourne, VIC 3207, Australia

314-321, 3rd Floor, Plot 3, Splendor Forum, Jasola District Centre, New Delhi - 110025, India

79 Anson Road, #06-04/06, Singapore 079906

Cambridge University Press is part of the University of Cambridge.

It furthers the University's mission by disseminating knowledge in the pursuit of education, learning and research at the highest international levels of excellence.

www.cambridge.org
Information on this title: www.cambridge.org/9781107658158

First published 2014

A catalogue record for this publication is available from the British Library

ISBN 978-1-107-05017-4 Hardback
ISBN 978-1-107-65815-8 Paperback

In memory of the first reader of this book, 1960
Paul Whittle 1938–2009

For mathematics is after all an anthropological phenomenon.

Wittgenstein (1978: 399)

Mathematical activity is human activity ... But mathematical activity produces mathematics. Mathematics, this product of human activity, 'alienates itself' from the human activity which has been producing it. It becomes a living, growing organism.

(Lakatos 1976: 146)

The birth of mathematics can also be regarded as the discovery of a capacity of the human mind, or of human thought – hence its tremendous importance for philosophy: it is surely significant that, in the semilegendary intellectual tradition of the Greeks, Thales is named both as the earliest of the philosophers and the first prover of geometric theorems.

(Stein 1988: 238)

A square can be dissected into finitely many unequal squares, but a cube cannot be dissected into finitely many unequal cubes. *Proof of the latter:*

In a square dissection the smallest square is not at an edge (for obvious reasons). Suppose now a cube dissection does exist. The cubes standing on the bottom face induce a square dissection at that face, and the smallest of the cubes on that face stands on an internal square. The top face of this cube is enclosed by walls; cubes must stand on this top face; take the smallest – the process continues indefinitely.

(Littlewood 1953: 8)

Contents

Foreword

This is a book of philosophical thoughts about proofs, applications, and other mathematical activities.

Philosophers tend to emphasize mathematical 'knowledge', but as G. H. Hardy said on the first page of his *Apology* (1940), 'the function of a mathematician is to *do* something, to prove new theorems, to add to mathematics'. I have emphasized the 'do'. Hardy was writing not only an *Apologia pro vita sua*, but also a mathematician's *Lament* that he was now too old to create much more mathematics. He also, notoriously, wanted to keep mathematics pure, whereas I believe that the uses, 'the applications', are as important as the theorems proved. Neither proof nor application is, however, as clear and distinct an idea as might be hoped.

To reflect on the doing of mathematics, on mathematics as activity, is not to practise the sociology of mathematics. Happily that is now a burgeoning field, from which one can learn much, but what follows is philosophizing, moved by old-fashioned questions – to which I add my title question, why *do* these questions arise perennially, from Plato to the present day?

This book began as the René Descartes Lectures at Tilburg University in the Netherlands, in early October, 2010. (I started writing out the talks on the summer solstice of that year.) The format was three lectures, each followed by comments from two different scholars. The original intention was that the lectures and comments would be published immediately.

I began to realize at the end of the week the extent to which the material needed to mature. The commentators generously agreed to keep their comments. So my first duty is to thank them deeply for their hard work. Hard work? Typically they received, late in the day, some 20,000 words per lecture, of which only 7,000 would be spoken, and they did not even know which ones.

For the first talk, 'Why Is There Philosophy of Mathematics?': Mary Leng and Hannes Leitgeb.

xiii

For the second talk, 'Meaning and Necessity – and Proof': James Conant and Martin Kusch.
For the third talk, 'Roots of Mathematical Reasoning': Marcus Giaquinto and Pierre Jacob.

Thank you all.

I originally proposed 'Proof' as the series title. That was the title of a thesis, which, together with some work in modal logic, was awarded a PhD by Cambridge University in 1962. It was dominated by my reading of Wittgenstein's recently published *Remarks on the Foundations of Mathematics*, although much influenced by what was to become Imre Lakatos' *Proofs and Refutations*, which was being completed in Cambridge as a doctoral dissertation when I began mine.

I have published very little about the philosophy of mathematics, but it has always been at the back of my mind, so the Descartes Lectures were a chance to finish the job. The title 'Proof' would give no idea of what the talks would be about, so Stephan Hartmann, the organizer of the events (to whom many thanks), and I hit on 'Proof, Calculation, Intuition, and A Priori Knowledge'.

Very soon after the Descartes Lectures, in late October 2010, I gave three similar talks at the University of California, Berkeley, beginning with the Howison Lecture, 'Proof, Truth, Hands and Mind'. Here is how I explained the title, after indulgently admiring my choice of words of one syllable:

> Why this title? First, because *proof* has been an essential part of Western mathematics ever since Plato. And Plato thought that mathematics was the sure guide to *truth*. I want also to think of how we do mathematics, in a material way that Plato would hardly have acknowledged. We think with our *hands*, our whole bodies. We communicate with one another not only by talking and writing but also by gesticulating. If I am thinking mathematically I may draw a diagram to take you through a series of thoughts, and in this way pass my thoughts in my *mind* over to yours.

After California I put this material aside while teaching on other topics at the University of Cape Town, and intensely experiencing all too little of that amazing land and its peoples. In January 2011 I did attend the annual meetings of the Philosophical Society of Southern Africa, and the corresponding Society for the Philosophy of Science, near Durban. There I presented, respectively, abridged forms of the first two Descartes/Howison Lectures (Hacking, 2011a, 2011b). I may mention also a contribution to a conference in Israel in honour of Mark Steiner, in December 2011, which began with

Pythagoras and ended with P. A. M. Dirac (Hacking, 2012b). Then in November 2012 I did part of the third Descartes Lecture as the Henry Myers Lecture for the Royal Anthropological Institute, London.

In March and April of 2012 I gave six Gaos Lectures at the National Autonomous University of Mexico, at the invitation of Carlos Lópes Beltrán and Sergio Martinez, to whom again many thanks. The title was *The Mathematical Animal*, but in fact the first five lectures covered only the first Descartes Lecture. And so it has come to pass that this book is not the entire set of lectures given in Tilburg, but only the first.

The connection between the present book and my dissertation of 1962 will not be obvious, but *plus ça change*. My title here is, *Why Is There Philosophy of Mathematics At All?* I was astonished, in preparing the present book for the press, to reread the brief preface to my dissertation of 1962: 'We must return to simple instances to see what is surprising, to discover, in fact, why there are philosophies of mathematics at all.' And I may mention that my choice of topics comes from the first edition of Wittgenstein's *Remarks on the Foundations of Mathematics* (1956). The two significant nouns most often used in that edition (to which I prepared my own index) are *Beweis* and *Anwendung*, 'proof' and 'application'.

I thank the Social Science and Humanities Research Council of Canada for awarding me its annual Gold Medal for Research. The cash coming with the medal is rightly dedicated to further research, and much of it was used in preparing this book. I thank James Davies in Toronto and Kaave Lajevardi in Teheran for a lot of help in the home stretch. The final threads were tied up in March 2013 during a blissful time at the Stellenbosch Institute for Advanced Study.

CHAPTER I

A cartesian introduction

1 Proofs, applications, and other mathematical activities

Why *is* there a whole field of inquiry, a discipline if you like, called the philosophy of mathematics? This unusual question, the very title of this book, will not begin to be examined with care until Chapter 3, but two summary answers can be stated at once.

First, because of the *experience* of some demonstrative proofs, the experience of proving to one's complete satisfaction some new and often unlikely fact. Or simply experiencing the power and conviction conveyed by a good proof that one is taught, that one reads, or has explained to one. How can mere words, mere ideas, sometimes mere pictures, have those effects?

Second, because of the richness of applications of mathematics, often derived by thinking at a desk and toying with a pencil. Or more poetically, in the words of the historian of science A. C. Crombie (1994 1, ix), 'the enigmatic matching of nature with mathematics and of mathematics by nature'.

Thus this book is a series of philosophical thoughts about proofs, applications, and other mathematical activities.

In line with the authors of my second and third epigraphs, Lakatos and Stein, I think of proving and using mathematics as activities, not as static done deeds. But at once a word of caution. One does not need proofs to think mathematically: it is a contingent historical fact that proof is the present (and was once the Euclidean) gold standard of mathematicians. Our common distinction between pure mathematics and applications is likewise by no means inevitable. (These assertions to be established in Chapters 4 and 5.) Thus the philosophical difficulties prompted by proofs and applications have arisen because of the historical trajectory of mathematical practice, and are in no sense 'essential' to the subject. What are often presented as very clear ideas, namely proof and application, turn out to be much more fluid than one might have imagined.

Two figures haunt the philosophy of mathematics: Plato and Kant. Plato, as I read him, was bowled over by the experience of demonstrative proof. Kant, as I understand him, crafted a major part of his philosophy to account for the enigmatic ability of mathematics, seemingly the product of human reason, so perfectly to describe the natural world. Thus Plato inaugurated philosophizing about mathematics, and Kant created a whole new problematic. Another version of my answer to the question, 'Why philosophy of mathematics?' is, therefore, 'Plato and Kant'.

A third figure haunts my own philosophical thinking about mathematics: Wittgenstein. I bought my copy of the *Remarks on the Foundation of Mathematics* on 6 April 1959, and have been infatuated ever since. What follows in this book is not what is called 'Wittgensteinian', but he hovers in the background. Hence it is well to begin with a warning.

2   On jargon

My first epigraph, 'For mathematics is after all an anthropological phenomenon', is taken out of context from a much longer paragraph, some of which is quoted in Chapter 2, §13 (henceforth, §2.13). The long paragraph is itself part of an internal dialogue stretching over a couple of pages. The words I used for the epigraph sound good, but bear in mind the last recorded sentence of Wittgenstein's 1939 *Lectures on the Foundations of Mathematics* (1976: 293): 'The seed I'm most likely to sow is a certain jargon.'

A good prediction! Recall some jargon derived from his words:

> Language game
> Form of life
> Family resemblance
> Rule following considerations
> Surveyable/perspicuous
> Hardness of the logical 'must'
> etc.... and:
> Anthropological phenomenon
> What we are supplying are remarks on the natural history of mankind.

Not to mention things he did not exactly say – 'Don't ask for the meaning, ask for the use.' (For what he did say, see §6.23.)

These are wonderful phrases. But they should be treated with caution and read in context. They all too easily invite the feeling of understanding. They are often cited as if they were at the end, not in the middle, of a series of thoughts. This book will quote scraps from Wittgenstein quite often. Hence it is prudent to begin with this *caveat emptor*, lest I forget.

Another notice to start with. I shall often allude to one or another of Wittgenstein's observations. But I shall not mention his thoughts about infinity, intuitionism, Cantor, or Gödel. And I will not discuss his ideas about following a rule, for a reason to be stated in §3.24.

3 Descartes

Descartes also created a certain jargon. These two exceptional philosophers may have far more in common than is usually noticed (Hacking 1982). Be that as it may, what follows began with adolescent infatuation with Wittgenstein, and its present form began as the René Descartes Lectures at the University of Tilburg.

The title of this biennial series of lectures is honorific, recalling only that Descartes (1596–1650) chose to live and work in the free intellectual climate of the Netherlands. There, between 1628 and 1649, he created a philosophy that would dominate the modern age. I was under no obligation to pay homage to the historical Descartes, but two distinct Cartesian themes will recur in what follows.

The first is the application of arithmetic to geometry – a central example of the application of mathematics to mathematics, itself a case of the application of mathematics to the sciences (including mathematics). This step in the great chain of applications, namely the application of arithmetic to geometry, is of particular note.

The second theme is the extraordinarily different conceptions of proof that seem to have been held by Descartes on the one hand, and, a little later, by Leibniz on the other. This is an extreme case of the great variety in types of proof that are used by mathematicians. This first modern confrontation between two ideal visions of proof foreshadows topics much discussed today, namely the impact of fast computation on the practice of 'pure' mathematics. Perhaps more important is a question that has newly become vital, when not long ago we thought we knew the answer: what *is* a mathematical proof?

Both topics are examples of later developments in this book. Neither theme is commonly thought of as Cartesian. Neither plays much of a role in contemporary philosophizing about mathematics, or in Cartesian scholarship. Thus this first 'Cartesian' chapter launches us into relatively uncharted territory, which will later be discussed in parallel with the well-mapped regions that today's philosopher expects to enter.

If there is any 'methodological' idea in this book, it is this: in philosophical thinking about mathematics, it is good to start by grasping straws from the tangled nest that is mathematics, straws that have not been much picked over recently. In the course of this chapter I shall also mention a number of other topics that will come up later, but which have received very little, if any, attention from philosophers of mathematics, although they certainly interest mathematicians. So this chapter comes, not in lieu of an introduction, but as a working introduction.

Since the two themes, of proof and application, appear to be unrelated, this chapter is divided into two parts. I say *appear*, for Wittgenstein may have thought differently: 'I should like to say: mathematics is a MOTLEY of techniques of proof. – And upon this is based its manifold applicability and its importance' (Wittgenstein 1978 (henceforth *RFM*): III, §46, 176). As already noted in my Foreword, the two most common substantive nouns (and their cognates) in the first edition of the *Remarks* (1956) were 'proof' (*Beweis*) and 'application' (*Anwendung*). But beware of jargon!

A APPLICATION

4 Arithmetic applied to geometry

> I have a vivid and happy memory of my first reading of Descartes, for it was with unbounded enthusiasm that I devoured the *Discourse on Method*, sitting in the shade of a tree in the Borghese Gardens on Rome in the summer of 1970. (Gaukroger 1995: vii)

People have been having this sort of experience since 1637, when the *Discourse* was published. Its full title is the *Discourse on the Method of Properly Conducting One's Reason and Seeking the Truth in the Sciences*. We tend to think of it as a free-standing work, and forget that it was published as the preface to three scientific treatises, the *Meteors*, the *Dioptrics*, and the *Geometry*.

The *Meteors* is about 'meteorology', understood as ranging from vapours to the rainbow. It is a perfect example of what T.S. Kuhn called a pre-paradigmatic discussion. Later he withdrew the idea of science before paradigms, but he was still pointing at something important in the study of some group of phenomena (Kuhn 1977: 295, n. 4). Kuhn's own example was *heat* at the time of Francis Bacon. In the case of Bacon, the heat from rotting manure is listed alongside the heat of the sun. Likewise in the *Meteors*, a vast range of phenomena are grouped together with no clear idea (for the modern reader) of what on earth they have to do with each other, except that they are above the earth but not, apparently, in the heavens.

The *Dioptrics*, about the theory of light, is in a much more advanced state, for much of it is about geometrical optics, followed by the theory of reflection and refraction, to which Descartes made important contributions, not always correct. It concludes with a physico-physiological discussion of how the eye works.

The *Geometry* is a thoroughly modern work, a surprising amount of which can be read by an interested reader without much difficulty today. I like to think of this third essay as one model of how to conduct one's reason and seek truth in a science.

I am not about to embark on an exposition of Cartesian philosophy, neither that of 'Mr Cogito' or 'Mr Dualist' on the one hand, nor, on the other hand, Descartes the mechanical philosopher, so brilliantly described by Martial Guéroult (1953) and then Daniel Garber (1992). What follows is one very thin slice from a very rich cake.

5   Descartes' *Geometry*

The *Geometry* made plain, for all the world to see, that there are profound relations between geometry and arithmetic. We now find it more natural to say that Descartes applied algebra to geometry, but Descartes did not see things in quite that way. He began by comparing arithmetical operations to geometrical constructions, and after listing many bits of arithmetical terminology, said that he would 'not hesitate to introduce these arithmetical terms into geometry for the sake of greater clearness' (Descartes 1954: 5).

Connections between geometry and arithmetic did not originate with Descartes. Many before him grappled with the same issues. The usual suspects are, in France, François Viète (1540–1603), and in Germany, the Jesuit astronomer and mathematician Christopher Clavius (1538–1612). In Italy we might go less to the university mathematicians than to the abacus tradition: that is, the work of men of commerce who actually used arithmetic, rather than the great innovators such as Cardano and Tartaglia who solved cubic equations and the like.

The eminent historian of science, John Heilbron, included in his *Galileo* (2010) a striking 'rational reconstruction' of Galileo's thought. It is an amazing dialogue in which 'Galileo' tries not only to put algebra together with geometry, but also to express time in such a way that it enters into geometrical/algebraic presentations. Minkowski portended?

Probably the history of arithmetic applied to geometry will go back further than the Italians, to the House of Wisdom in Baghdad, where algebra as we know it matured in the tenth century, and probably to Pappus of Alexandria (about 290–350). Descartes did not come from nowhere but from everywhere. He owed far more of his geometrical results, about, for example, conic sections, to Apollonius (262–190 BCE) than he cared to acknowledge. Yet it was he who turned all these explorations into a perspicuous whole that novices could master.

6 An astonishing identity

Many difficult problems in geometry can be solved by turning them into arithmetic and algebra, and many problems in the theory of numbers and algebra can be solved by turning them into geometry. This continues, as we shall soon illustrate, from the time of Descartes to the present day. It is as if geometry and the theory of numbers turn out to be about the same stuff. I find this astonishing.

In contrast most people take it for granted if they think about it at all. One scholar whom I greatly respect, and to whom I expressed my concern, said, 'it is obviously over-determined'. Maybe. But let us begin modestly, thinking only in terms of the application of one science to another, and in particular of Descartes 'applying' arithmetic and algebra to plane and later solid geometry.

7 Unreasonable effectiveness

Algebra, born of arithmetic, can be applied to geometry. That is the first indubitable example of the unreasonable effectiveness of mathematics developed for one purpose, applied to mathematics developed for another purpose. (Others, more knowledgeable than I, might cite Archimedes over Descartes.)

My tag, 'unreasonable effectiveness', is a deliberate play on the title of a famous essay by Eugene Wigner: 'The unreasonable effectiveness of mathematics in the natural sciences' (1960). Wigner's phrase has become a cliché. He was surprised that quite often 'pure mathematics' developed, as Wigner put it, primarily for aesthetic purposes, has turned out to be an invaluable tool in the sciences. The most familiar example – perhaps suspiciously over-used – is the way in which Riemann's non-Euclidean geometry was precisely the tool needed for special relativity.

Wigner's question has been asked often, although seldom with the persistence of Mark Steiner's *The Applicability of Mathematics as a Philosophical Problem* (1998). We shall turn to these issues later. Descartes reminds us that there is another question which is seldom addressed. How come mathematics developed in one domain, with one set of interests, turns out to make a critical difference to mathematics developed in another domain, with an apparently different subject matter? My play on Wigner's 'unreasonable effectiveness' of mathematics, applied not to physics but to mathematics, has already been made by Corfield (2003).

Why should the application of algebra to geometry be 'unreasonable', or at least unexpected? Here are three reasons to start with.

According to the traditional history of 'Western' mathematics, geometry is Greek, and arithmetic and algebra start in India and pass through Persia and Islam. On that account, they have distinct historical origins. Second, Kant plausibly made arithmetic out to be the synthetic a priori truths of time, while geometry constitutes the synthetic a priori truths of space. Third, much contemporary cognitive science locates the 'number sense' in parts of the brain distinct from those deployed in spatial sensibility.

It has become something of a truism that people in most societies tend to imagine the series of numbers as arranged on a line. Remember, however, most truisms, if true at all, are at best 'true for the most part'. Núñez (2011) argues that this linear representation of the numbers is recent even in the West, and that there is no trace of it in Babylonian mathematics or among many Amazonian peoples. A more interesting thought is about Mayan civilization. The world did not come to an end at the end of 2012, as some ludicrous misreadings of Mayan calendars led the fanciful to fear, but the Maya may well have had a circular sense of number rather than a linear one. But of course that is only another geometrical, spatial, representation of number.

The interconnections between geometry and algebra are, nevertheless, almost too good to be true. Geometry is spatial, and algebra is a child of arithmetic. Arithmetic is for counting, a process that, as Kant emphasized, takes place in time. I don't mean that you cannot do arithmetic without time, or know cardinalities without counting. You can usually tell just by looking that a group has exactly four or five or maybe even six members. (That's called 'subitizing': infants can discriminate between groups of two and three. So can monkeys, pigeons, crows, and dolphins.) After that, you have to count to be sure. Counting, Kant would have insisted, progresses in time, and counting is almost certainly not innate but had to be learned. But you can often tell, just by looking, that one group of things is larger than another. You can tell that two sets have the same number of members by matching them, like Frege's waiter (Frege 1952: 81). Nevertheless, I do think that Kant's initial location of arithmetic in time is important, and brings out a difference between it and geometry in space.

Although constructing a figure or even drawing a line takes time, Kant was right to say you can see a circle is a circle just by looking, and even if you can't name an irregular shape, you can see it is a shape. Space and time, manifested in geometry and arithmetic, are the twin poles of Kant's Transcendental Aesthetic. (Why 'Aesthetic'? Because the aesthetic deals with what he called sensibility, not reason.) He used space and time, geometry and arithmetic, as the road to an entire philosophy that has haunted us ever since.

All right, it will be protested, Kant did think that arithmetic and geometry are fundamentally different, but that part of his philosophy has not stood up very well. Most would say that his ideas about geometry were refuted by special relativity in 1905; his views about space and time were refuted by general relativity in 1916; his views about causation were refuted by the second wave of quantum mechanics in 1926. Early in the twentieth century L. E. J. Brouwer (1881–1966) did found Intuitionism on Kantian principles about continuing sequences in time, but his chief legacy seems to be constructive mathematics, which seldom makes use of Kant's motivating thought. So why should we, today, be impressed by the well-known fact that arithmetic can be applied to geometry?

One answer, as already suggested, derives from contemporary cognitive science. There is the now popular modular view of the brain and cognition. It is a plausible conjecture that the modules that enable us to navigate our spatial environment, and invite geometry, are distinct from those that give us *The Number Sense*, to use Stanislas Dehaene's title (1997). Figuratively speaking, we have two distinct cognitive systems, each run by its own neural network in the brain. Very plausible. (Just the sort of thing that Kant might have expected!) There is at least something of a consensus, at present, that there are distinct 'core systems', one group of which enables judgements of numerosity, and another of which enables simple geometrical judgements. Yet somehow the two, arithmetic and geometry, turn out to be about the same stuff.

8   The application of geometry to arithmetic

Descartes applied arithmetic to geometry. The tables can be turned, and geometry applied to numbers. Let us take, for example, a name already mentioned, that of Hermann Minkowski (1864–1909). He was one of Einstein's teachers, and the better mathematician. It was Minkowski who, in 1907, realized that the special theory of relativity should be conceptualized in four-dimensional space-time. In 1908 he gave a fundamental lecture to the annual meeting of German scientists and physicians; it began with the rousing words:

> Henceforth space by itself, and time by itself, are doomed to fade away into mere shadows, and only a kind of union of the two will preserve an independent reality. (Minkowski 1909)

Kant on space and time demolished in a sentence? Maybe. But Minkowski is relevant here for quite a different reason. He was first famed as an extraordinary innovator in number theory. Yet he made most of his discoveries in number theory – and then in mathematical physics and in the theory of relativity – by using geometry. (See Galison 1979, a fascinating paper he wrote as an undergraduate.) That is the inverse of the Cartesian application of algebra to geometry.

9   The application of mathematics to mathematics

The application of geometry to number theory, and of algebra to geometry, are special cases of the application of mathematics to mathematics. Because of my concern with the proposed distinct cognitive origins of geometry and numbers, I find it deeply perplexing. The more general topic of the application of one bit of mathematics to another bit of mathematics has received curiously little attention from philosophers. It is a constant source of both delight and achievement among mathematicians. Many of the examples of recent mathematics that I shall use in what follows take an idea from one branch of mathematics and use it to create or to solve a fundamental problem in another.

Kenneth Manders is the rare philosopher who has taken the topic very seriously. 'We are accustomed to the idea that something *prima facie* intellectually powerful is happening in empirical applications of mathematics, but philosophers . . . have yet to admit a parallel in the math-math world.' I quote from 'Why apply math?', an unpublished paper written in the 1990s. His title is at first glance misleading, for it suggests an interest in empirical applications, but which were precisely not his target. Yet the title is not inapt, for he thinks that grounds for the applicability of mathematics may 'lie in features shared by math-empirical and math-math applications alike'. I share his suspicion.

Manders explains philosophical indifference to 'math-math' by the twentieth-century attitude that mathematics is all deduction, all proofs. 'From this point of view, there is indeed nothing special about applying a mathematical theorem in a mathematical proof: it's just proofs and more proofs.' As illustrations of this attitude he cites a letter of Hilbert's to Frege (21 February 1899; Frege 1980) and a famous paper by C. G. Hempel, first published in 1945 in the *American Mathematical Monthly*. It is clear that Hempel believed that the application of mathematics to any field was a matter of interpreting a set of axioms in a new context: what Manders calls the 'interpretive' conception of the application of mathematics. I am tempted to go further: after Gödel and Tarski, logicians and philosophers were justifiably obsessed with a semantic approach to mathematics. That deflected attention from what is being done when one domain of mathematics is applied to another. To paraphrase Manders, it made it look like models and more models.

The idea identified by Manders, and which he says has misled us, and which I connect with the semantic approach, is of an abstract formal system within which one makes valid deductions; then there are various interpretations of the system. A body of mathematics is applied to a subject matter when the subject matter provides an interpretation for the mathematics. All deductions valid in the one lead, under the interpretation, to sound conclusions in the other. And that's all there is to applying one branch of mathematics to another. Manders observes that things are not like that at all.

He takes as his example of 'math-math' application not Descartes, but a theorem discovered by Girard Desargues (1591–1661) in 1647, ten years after Descartes had published his *Geometry*. The result lay dormant for a long time, until the revival of projective geometry in the nineteenth century. Manders argues that the resulting reconception of Euclidean geometry is not just a matter of reinterpretation of axioms, but a change in the ways in which space can be conceived and presented to the mind. Projective geometry (to simplify Manders' subtle paper) was applied to Euclidean geometry, and our understanding of both was affected by the interchange. (A little more about projective geometry in §§4.5 and 5.11.) The arid one-way tale of axioms and interpretations completely ignores the dynamics of the situation, or so Manders argues. This is an important insight, which is becoming increasingly widely shared. It will be illustrated in detail in connection with the 'pure' and 'applied' mathematics of rigidity in §§5.24–26.

The theme, of how technical inventions can alter human possibilities for thinking about space, is itself worthy of philosophical thought. That is what Hermann Minkowski did with his diagrams, and, in the 1960s, what Roger Penrose and Brandon Carter did with conformal diagrams, now often called Penrose diagrams. With a lot of training, a very clever young person can learn to represent the causal properties of infinite dimensional space on a flat piece of paper or a whiteboard. And then to think productively about black holes and many things even more esoteric. That is what Stephen Hawking does, but how does he do it when for so long he has been unable to draw a diagram? Hélène Mialet (2012) describes how every year he chooses a top entering doctoral student, with whose advanced diploma work he is familiar. The student is assigned a research topic within Hawking's interests, and learns to draw the diagrams Hawking needs. The student is working for him for four years, so at any time there will be four 'prostheses' – students enabling Hawking to think.

10   The same stuff?

A platonist, who reads the historical Plato in one of many possible ways, might say that geometry and number theory turn out to be the same stuff, the same mathematical reality that can be approached, in blinkered human fashion, beginning with arithmetic and plane geometry. Likewise, philosophers or mathematicians who say that the essence of mathematics is neither the objects it studies, nor the theorems it proves, but is instead the study of structure, will conclude that Descartes established once and for all that geometrical structures are closely related to or identical with algebraic structures. Charles Chihara's 'New directions for nominalist philosophers of mathematics' (2010) explicitly invokes the application of mathematics to mathematics to argue exactly that. So 'platonist' and 'nominalist' can take math-math applications to argue for their theses. That seems surprising, for these philosophical schools – platonist, nominalist, structuralist – are supposed to be at odds. 'Platonism', 'nominalism', 'structuralism': I shall put aside these less than clear labels for as long as I can.

The logician and philosopher Paul Bernays, thinking, in a 1934 lecture, in a Kant-derived way, seemed to conclude that, in my crude form of words, geometry and arithmetic are *not* what I call the same stuff (Bernays 1983: 268–9). He said that 'in geometry the platonistic idea of space is primordial'. In contrast, 'the idea of the totality of numbers is superimposed' on arithmetic, and so number theory appeals to the intuitionist 'tendency'. There is at most a 'duality of arithmetic and geometry [that] is not unrelated to the opposition between intuitionism and platonism' (1983: 269). A little more on this in §7.6 below.

The Kantian distinction between arithmetic and geometry was a given for those educated in a neo-Kantian or phenomenological tradition – and that includes both Brouwer and Hilbert, as well as Bernays – so that the applicability of the one to the other was more striking than for empiricists and logicists.

11   Over-determined?

Perhaps my astonishment is unwarranted. From earliest times geometry has been concerned with measurement, which is numerical. How could architects have proceeded otherwise? In the King James Bible, we read that Hiram, the Phoenician artisan employed by Solomon on the building of the Temple, 'made a molten sea [of brass] of ten cubits from brim to brim, round in compass, and five cubits the height thereof; and a line of thirty cubits did compass it round about' (2 Chronicles 4.2; cf. 1 Kings 7.23). That suggests that the authors believed that the value of what we call π is exactly 3. Apologists reject this reading, but no matter what the original text meant, the passage indicates an awareness that there was a ratio (arithmetical) between the circumference of a circle and its diameter (geometrical).

For another example from another civilization, the cult of Pythagoras was deeply impressed with harmonics. We start with the string of a lyre, a geometrical object, namely a straight line. (Only later do Neoplatonists distinguish between the ideal object, the straight line, and the physical taut string.) We observe (or rather hear) that the string of a lyre (for example) can be divided in various ratios to produce the octaves. A harmonic progression is defined as a sequence $x/1$, $x/(1+d)$, $x/(1+2d)$, ... Archytas (428–347 BCE), an exact contemporary of Plato and his chief informant about Pythagorean reasoning, (may have) discovered the harmonic progression: 1, 2/3, 1/2 ... [i.e. 2/2, 2/3, 2/4 ...], which corresponds to the heard harmonics of a plucked string.

The string is a line (geometrical), which, plucked at various ratios (arithmetical), gives sounds that agree with the ear. There are many more complex connections between arithmetic series and heard harmonics, which the Pythagoreans made into an entire cosmology – 'the music of the spheres'. It all begins by putting arithmetic – not whole numbers but ratios between whole numbers – together with the properties of the line. Hence the Pythagorean dream: the universe just *is* mathematical in character. This idea is still going strong with some mathematical physicists. 'Our physical world *is* an abstract mathematical structure' (Tegmark 2008: 101, italics added, to emphasize that he is saying the universe is such a structure, and not merely that it has such a structure). That is Tegmark's Mathematical Universe Hypothesis. More modestly, Dirac asserted only that there must be 'some *mathematical quality in Nature*, a quality which the casual observer of Nature would not suspect, but which nevertheless plays an important rôle in Nature's scheme' (Dirac 1939: 122, original capitalization and italics).

The Pythagorean dream – or fantasy – has not much interested philosophers of late, with the exception of Steiner (1998). I take it up briefly in §§4.8–11, and at length in Hacking (2012b). Although I classify the dream as fantasy, I shall suggest that the very idea of mathematical physics, that mathematics can be used to unlock the deepest Secrets of Nature, owes a great deal to a Pythagorean tradition.

We know little about the historical Pythagoras, but a fair amount about the Pythagorean cult, for which Kahn (2001) is the best short guide. We have the phrase, 'Pythagoras' theorem'. The area of the square on the hypotenuse of a right triangle is equal in area to the sum of the areas on the other two sides: I shall call that the Pythagorean *fact*. One thing we do know is that the fact was familiar before Greeks even turned their hand to mathematical reasoning. Perhaps members of the cult were the first to prove it, in something like the form of Euclid x.9. Hardy (1940) takes the theorem to be about numbers; in the special case, $\sqrt{2}$ is 'irrational'. Did Archytas or whoever see things that way? I doubt it, but if he did, that is just another instance of the primordial sensibility that numbers and shapes are intimately related.

Isn't the intertwining of number with geometry over-determined? Even the Euclidean result most familiar to readers of Plato – the doubling of the area of a given square in *Meno*, often called a case of Pythagoras' theorem – involves doubling, or 'two times'. We still have a congealed memory of the idea when in English (and a number of other languages) many of us learned to say, colloquially, 'four times seven makes twenty-eight'. Note the *times*. That goes right back to moving a length made of four identical 'unit' lengths end to end, and continuing in a straight line seven times; the result is twenty-eight lengths. That takes time. Children in my childhood still called it the 'times table' – time and numbers and geometrical lengths all tied up in primary school.

12 Unity behind diversity

The history of mathematics is one of diversification and unification. We start with diversity. Some peoples, namely the ancient Greeks, fixed on geometry as primary, while others, in India and working in Sanskrit, fixed on numbers as primary, and bequeathed that obsession to Islam and Arabic. But it all keeps on turning out to be the same stuff!

For those who read science journalism, the most widely reported mathematical breakthrough in recent times was made by Andrew Wiles. He 'applied' abstruse features of elliptic functions to a home truth of arithmetic, Fermat's last theorem. It is a source of mathematical joy that structures developed for one purpose, at first following one barely visible track through the woods, should cross a well-known path and suddenly solve its problems.

Many examples of unification of the diverse are so long ago that we have forgotten them. Recall, for instance, the fact that we needed the complex plane to prove some elementary facts of arithmetic. Imaginary numbers are called in to sort out the whole numbers. Any reader familiar with some mathematical field of specialization – no matter what – will know how it has been advanced by work in another field that started from unrelated insights and motives.

To say that complex numbers have been needed for the number-theoretic discoveries does not mean that they will always be deemed essential for proof. It is a constant urge among one type of mathematician, including major proof theorists, to reduce proofs to a minimum of assumptions. Thus Atle Selberg (1917–2007) was awarded the Fields Medal (1950) in part for finding the first elementary proof of the prime number theorem.

That is the theorem that tells about the distribution of the primes among the integers – that it is essentially random. No matter how we select subsets of prime numbers, their arrangement will be just as if the numbers were coming off a roulette wheel. This is one half of a pair of facts which creates what we might call the prime numbers conundrum. On the one hand, they appear haphazard in the series of integers. On the other hand, there are enormously powerful laws that apply to all primes.

Deriving a proof of the prime number theorem that did not go outside the theory of whole numbers was rightly regarded as a remarkable achievement – so remarkable that there was a nasty priority dispute between Selberg and Paul Erdös (1913–96) about who found out what. (See Spencer and Graham 2009, but note that Graham was a close collaborator of Erdös.) It was a beautiful case of reducing the assumptions needed to prove a fundamental theorem.

Michael Atiyah (b. 1929, Fields Medal, 1966) is not untypical in his fascination with surprising unifications:

> But I like two things in mathematics – unification, things that unite, unexpectedly, you know. Many beautiful things in mathematics prove something in subject A. By a marvellous and unexpected link a subject way in the corner over there, if you apply this idea then presto, you get to solve those problems, and that's the sort of thing I like. That's one form of unexpected thing. In general, the unexpected. (Atiyah 1984)

13   On mentioning honours – the Fields Medals

So Selberg got a Fields Medal for his proof: why mention that? I do not like citing medals of recognition, for they suggest an appeal to authority. Some and perhaps many readers will, however, not be familiar with the names of mathematicians whose opinions I mention. Hence I shall note some who have been recognized in their community (and hence can readily be Googled). The Fields Medal is the usual name for the *International Medal for Outstanding Discoveries in Mathematics*. Fields Medals are awarded every four years to at least two and at most four mathematicians under forty years old. They were established in 1936 by an otherwise forgotten Toronto mathematician after whom they (and the Fields Institute for Research in Mathematical Sciences) are named. They have a very modest cash prize, but confer enormous respect in the mathematics community.

There are many other prizes in mathematics, some of which carry a lot more money, and you can readily look them up. It is often observed that there is no Nobel Prize in mathematics, but it is less well known that there are other Nordic prizes comparable in prize money to half a Nobel (and since science Nobels are usually shared, de facto equivalent). There are the Crafoord Prizes awarded by the Swedish Academy for work on subjects that complement fields in which Nobels are awarded: thus, in 2012, to Jean Bourgain for work described as 'harmonic analysis, partial differential equations, ergodic theory, number theory, functional analysis and theoretical computer science'. Norway awards the equally valuable Abel Prize for lifetime achievement in mathematics.

Then there are the Millennium Prizes (§2.21), the Clay Research Awards, and the Wolf Prize; the Reid, von Kármán, and Wilkinson Prizes in applied maths; the list goes on. And on. But I shall mention only Fields Medals.

To repeat, I do not do so in order to confer authority on a mathematician's words. Historians of the sciences usually find that Grand Old Men are not particularly good at philosophical or historical insight into the scientific field for which they are famed. But some are, and it is of interest when they write from their experience as mathematicians, and not as trained philosophers. Chapter 6 plays out Platonism vs anti-Platonism by pitting the opinions of two articulate Fields Medallists against each other.

Moreover, although not all great mathematical innovations come from persons under forty years of age, there is some truth in the statement that the simple list of all Fields Medals goes a long way towards recording the notable mathematical discoveries and innovations of, say, the past sixty years. That is a paraphrase from a remark made by James Arthur of the

Arthur–Selberg trace formula – the same Atle Selberg as in §12 – when he was introducing, in a slightly institution-serving way, the Inaugural Fields Medal Symposium, 15–18 October 2012, at the Fields Institute in Toronto. The intense four-day symposium was dedicated to Ngô Bao Châo, Fields Medal 2010, for proving the Fundamental Lemma of the Langlands programme, to which we turn briefly in §15 below.

I shall indeed often mention a Fields Medal, but as an antidote to over-reverence it is useful to recall a remark by Robert Langlands himself:

> mathematics is a joint effort. The joint effort may be, as with the influence of one mathematician on those who follow, realized over time and between different generations – and it is this that seems to me the more edifying – but it may also be simultaneous, a result, for better or worse, of competition or cooperation. Both are instinctive and not always pernicious but they are also given at present too much encouragement: cooperation by nature of current financial support; competition by prizes and other attempts of mathematicians to draw attention to themselves and to mathematics. (Langlands 2010: 39f, n. 2)

14   Analogy – and André Weil 1940

Perhaps we should not speak so much of the applicability of arithmetic and algebra to geometry, as of analogies between them, and perhaps also – as Bernays did (quoted in §10 above) – of a duality between them. Here it may be helpful to turn to a letter that André Weil (1906–98) wrote to his sister, Simone Weil (1909–43), in 1940.

The Weils were remarkable siblings. André Weil, number theorist, algebraic geometer, and much else, had an immense influence on twentieth-century mathematics, not only as a leader of the Bourbaki group (§2.12). Simone Weil was one of the deepest philosophical moralists, often described as a mystic. With some justice, Albert Camus called her 'the only great spirit of our times' – that is *esprit*, which also means 'mind' (in a newspaper article of 11 February 1961, cited in Pierce (1966: 121)). The letter was written while André Weil was briefly imprisoned for trying to avoid military service in France at the beginning of World War II. Although he asserted that he was a conscientious objector, he relented, joined the army, and then absconded to the United States. Simone Weil, in the same period, was very active in the Resistance, but we shall pass by these divergent stories.

Weil's citation for a very large Japanese prize awarded in all fields of endeavour (the Kyoto Prize) states, in part, that he 'started the rapid advance of *algebraic geometry and number theory* by laying the foundations for abstract algebraic geometry and the modern theory of abelian varieties. His work on algebraic curves has influenced a wide variety of areas, including some outside mathematics, such as elementary particle physics and string theory' (*Notices of the American Mathematical Society* 41 (1994), 793–4). I have added italics to emphasize that not only did he advance both algebraic geometry and number theory, but also that he did so in part by establishing and developing analogies between the two. This is exactly what he describes in the 1940 letter to his sister.

The letter of 26 March 1940 describes his 'arithmetic/algebraic work', some events in the history of number theory, and 'the role of analogy in mathematical discovery' – 'examining a particular example, and perhaps you [Simone] will be able to profit from it' (Weil 2005: 355). He does not expect his sister to understand all or even most of the letter. It does indeed get into quite deep waters – which I shall not enter. I mention it here because Weil does not think of applying one field of mathematics to another, but instead of seeing analogies between the two, namely analogies in underlying structures. That led Weil and his fellow members of the Bourbaki group to think that mathematics is about structures (§2.12).

In passing I may mention that Weil's first example is what Gauss (1777–1855) called the 'Golden Theorem', or more descriptively, the law of quadratic reciprocity about prime numbers and the solution of polynomial equations. The law had been recognized by Legendre (1752–1853). In the first instance it took the young Gauss some nine months to prove it, publishing it in 1801; in the course of his life he published eight proofs. Nowadays it occurs early in any course on number theory. Courant and Robbins (1996), intended for beginners with nothing much past high school mathematics, gets there by page 37. Tappenden (2009: 259–63) has a useful philosophical analysis of why this result has fascinated mathematicians since the time of Legendre and Gauss. It is a beautiful example of one side of the conundrum mentioned in §12 above, of deep laws among the primes, running alongside the sheer 'randomness' with which primes occur among the integers.

In the letter to his sister, Weil begins with reciprocity and proceeds through a brief history of algebra, explaining how it has become clear to him that there are extraordinary analogies among seemingly unrelated fields of mathematics – fields whose origins have no historical connection with each other. Often, when stumped in one field, one turns to an analogical field in order to make progress. This is not so much a matter of applying one field to another, but of exploiting correspondences that slowly reveal themselves. The law of quadratic reciprocity has generated endless new mathematical problems, whole new classes of numbers. Tappenden (2009: n. 12) observes that 'a 2002 Fields medal was awarded [to Laurent Laforgue] for work on the Langlands programme, an even more ambitious generalization' of quadratic reciprocity. So what is this programme?

15   The Langlands programme

The discovery that arithmetic can be applied to geometry, and vice versa, must seem a long time ago – a closed book. Hence we should give a contemporary reminder. But then there is a problem, for an important example will have to be taken from a field so highly developed that it will be abstruse even to most mathematicians who do not work in that area. But if we do not show that there is something surprising, even today, then my question can be relegated to the dusty archives of the historian. A philosophy of mathematics should at least notice real and current mathematics. Better to be a science journalist than stuck in the mud. But also we have to be wary. Journalists are attracted to sensationalism. We should not rely on the 'glitter', as Wittgenstein called it, of overly sensational or simply exotic mathematics.

With that excuse, I shall now mention something I obviously do not understand.

As mentioned in §13, one of the 2010 Fields Medals was awarded to Ngô Bao Châo (b. 1972) for the proof of the 'Fundamental Lemma' of the Langlands programme. The programme has turned out to be incredibly rich, which means that it demands far more explaining that I am able to do. I introduce it to illustrate contemporary applications of number theory to geometry, and vice versa. Or is it better thought of as seeing the analogies between the two? Describing Ngô's result, Peter Sarnak, a distinguished mathematician close to these matters, is reported (in a broadsheet issued by the Institute of Advanced Study, Princeton) to have said,

> It is very rare that you can take a proof in the geometric setting and convert it to the genuine number theoretic setting. That is what has transpired through Ngô's achievement. Ngô has provided a bridge, and now everyone is using this bridge. What he has done is deep. It is below the surface and it is understanding something truly fundamental. (Institute for Advanced Study 2010: 4)

So here we have the same surprise that might have been felt early in the seventeenth century, when it was realized that you can apply arithmetic to geometry. But now we have some explaining to do.

The Langlands programme is named after work by Robert Langlands (b. 1936),*[1] starting with a now famous letter that he wrote to André Weil in 1967 (see Langlands website). Langlands sees his programme as growing almost seamlessly out of 'a continuation of developments in the theory – algebraic and analytic – of algebraic numbers created in the nineteenth and early twentieth century' (email of 24 August 2010). This body of work will not be easy even to refer to. For a glimpse of the difficulties, an early 'Elementary introduction to the Langlands program' says at the start:

> Herein lies the agony as well as the ecstasy of Langlands' program. To merely state the conjectures correctly requires much of the machinery of class field theory, the structure theory of algebraic groups, the representation theory of real and p-adic groups, and (at least) the language of algebraic geometry. In other words, though the promised rewards are great, the initiation process is forbidding. (Gelbart 1984: 178)

There is much more to the programme now, more than twenty-five years after those words were written. Whole new fields of research have been created. (The Arthurs–Selberg trace formula mentioned in §13 is one ingredient.) Still we may at least glimpse the way in which the programme is the working out of a dense set of analogies. On the one hand, there are structures that arise naturally in number theory, derived from roots of polynomial equations. (These are called Galois representations, acknowledging roots in the work of Évariste Galois (1811–32), the legendary pioneer of group theory.) On the other hand, there are structures that arise in geometry or mathematical physics, and which exhibit deep symmetries. (The structures in question are called automorphic forms, first glimpsed clearly by Henri Poincaré (1854–1912).)

In his letter to Weil, Langlands in effect proposed that there is a fundamental analogy between the two seemingly unrelated types of structure, and that one can derive results in the one field by working in the other field. Establishing special cases of these relationships has become a subfield of mathematical research. It may be that the best introduction to the Langlands programme, for those wholly ignorant of it, is a long essay that Langlands prepared for a philosophy conference discussing beauty – he was assigned beauty in mathematical theories (Langlands 2010). I think he was unusually sensitive, and careful, on 'aesthetics', but the interest of the essay is in his exposition of the history of mathematics, emphasizing events over the course of two centuries that lead up to his own work – 'almost

---

[1] A * indicates that a remark will be made in the 'Disclosures' section at the end of the book.

seamlessly', to repeat the words of his email. One may notice a certain parallel between this way of introducing ideas to a lay person and the way in which André Weil wrote to Simone Weil in 1940.

I have been emphasizing how Ngô's proof of the Fundamental Lemma of the programme illustrates the inter-applicability of number theory and geometry, as indicated by Peter Sarnak's observation quoted above. To call it a 'lemma' makes it sound just like a stepping-stone, and it was at first thought to be just that. Langlands introduced it in lectures given in Paris in 1979 (later printed as lecture notes, see Langlands website). Andrew Wiles is reported (in the same house broadsheet as Sarnak) as observing that it is now merely curious that it is called a lemma. 'It is a theorem. At first, it was thought to be a minor irritant, but it subsequently became clear that it was not a lemma but rather a central problem in the field' (Institute for Advance Study 2010: 1).

On the score of 'application', we should note that the Langlands programme has been taken up with enthusiasm in at least one physics community, with the eminent string theorist Edward Witten among those leading the way. (He was the first ever physicist to be awarded a Fields, in 1990.) He has co-written a book-length paper showing how the geometric Langlands programme makes deep sense of a nineteenth-century phenomenon, the duality of electricity and magnetism (Kapustin and Witten 2006). Here we have a handsome confirmation of Manders' suspicion, that the application of mathematics to physics is closely connected with the application of mathematics to mathematics. Indeed what is called the geometric Langlands programme originated with ideas that struck Vladimir Drinfel'd (b. 1954, Fields Medal 1990) as a teenager, and which much extended the scope of the Langlands correspondences between seemingly unconnected domains. Drinfel'd has gone on to make fundamental contributions to mathematical physics. Is this application? Or is it rather the working out of ever deeper analogies?

16   Application, analogy, correspondence

The trouble with the very notion of 'application' is that it is a one-way concept: we apply A to B. To counter this I have invented the word 'inter-applicability'. Mathematicians who work in these domains more often speak of correspondence, which is a symmetric notion. Although I began this discussion in terms of application, we might better have spoken of Descartes establishing a correspondence between arithmetic and geometry. I once in conversation spoke of Descartes arithmetizing geometry; the person with whom I was speaking, having in mind the way algebraic problems of the day could now be solved geometrically, observed that Descartes had geometrized algebra. Exactly so.

Weil's 'analogy' is likewise symmetric: if A is analogous to B, then B is analogous to A. Should we then revise Manders' terminology and speak not of math-math applications, but of math-math correspondences or analogies? We might then return talk of application to its usual site, applied mathematics, or the applications of mathematics outside mathematics. Some of Manders' critique of philosophers writing about application might then become moot. That is true of what I shall later call mission-oriented applied mathematics, where we do mathematics in order to solve problems in technoscience. That is often very much an asymmetric, one-way, type of reasoning, with the mathematics on one side and the applications on the other. But as I shall illustrate in Part B of Chapter 5, a great deal of applied mathematics involves a two-way exchange. One day mathematics is teaching about nature, while on another day, nature is teaching the mathematician.

B PROOF

17 Two visions of proof

Ever since the time of Euclid, proof has been the gold standard of mathematical achievement in the West. I do not believe that was inevitable. At any rate, it is far from a simple story. I shall sketch some of its twists and turns in Chapter 4. For the present we notice just one decisive fork in the history of proof, a contrast between two ideal conceptions, one expressed by Descartes, the other by Leibniz.

There are proofs that, after some reflection and study, one totally understands, and can get in one's mind 'all at once'. That's Descartes.

There are proofs in which every step is meticulously laid out, and can be checked, line by line, in a mechanical way. That's Leibniz.

18 A convention

I shall call these two ideals cartesian and leibnizian.

In English we tend to capitalize ideas and doctrines named after great men, Marxism, for example, and Marxist. In German, *Marxismus* is automatically capitalized as a noun, while *marxistische* is the usual form of the adjective. In French, *marxisme* and *marxiste* are the rule, both lower-case. English is more flexible. When a doctrine is connected by a strong historical tradition with a historical figure, I shall capitalize it: the Cartesian doctrine of two substances, mind and body. When the idea is not so closely connected, I shall not capitalize it; thus I speak of a cartesian ideal of proof.

Much later I shall speak of Platonism and Pythagoreanism, referring to ideas about mathematics that appear to be quite strongly connected with the historical Plato, or with the historical cult of the rather legendary Pythagoras. On the other hand, platonism and pythagoreanism will refer to contemporary ideas that remind us of the ancients, but have little in the way of strict historical connection to them. The distinction cannot be sharp, especially since Plato was a Pythagorean, and many modern pythagoreans are not only platonists but Platonists.

19   Eternal truths

My convention is useful in connection with proof. It is correct to say that two very different ideas of proof can be read into the obsessively careful writings of Descartes, and can be extracted more explicitly from some of Leibniz's innumerable texts. The two ideas became clearly separable only in the twentieth century, and that, as usual, is a complicated tangle of many local events converging on an emerging consensus.

I think it helpful to see that the outlines of the distinction are there from the beginning of recognizably modern mathematics, but it would be wrong to say that Descartes or even Leibniz saw matters clearly and distinctly. One of many things that divided the two philosophers was the nature and characteristics of truths known or established in mathematics. It was an old debate that much exercised anyone who believed, as mediaeval Christian and Muslim philosophers did, in an omnipotent and omniscient God. A square always has four sides, and 2+2 will always make 4: those are eternal truths. But could God, if he so chose, make a five-sided square, or make 2+2 make 5? If not, then there is something he cannot do, so he is not omnipotent.

This debate was remembered even in the *Tractatus* (3.031), where Wittgenstein wrote: 'It used to be said that God could create anything except what would be contrary to the laws of logic.' Well, what would stop him? Wittgenstein's answer at that time was that 'we could not *say* of an "unlogical" world how it would look'. (I have used the first, Ogden, translation (1922). The Pears and McGuinness translation (1961) is: 'We could not *say* what an "illogical" world would look like.')

20   Mere eternity as against necessity

Descartes was the rare modern philosopher who did think that God could make 2+2=5. The eternal truths will, in fact, be true forever, but they don't have to be that way. A modal logician distinguishes kinds of necessity, and might say that the eternal truth is necessarily true, but not necessarily necessary (S4 but not S5). I think the older conception may be more apt, not only for Descartes, but also for thinkers such as Bertrand Russell who thought all we need for clear thinking is the universal quantifier, which ranges over everything, and hence over eternity. No need to invoke necessity. More on Russell's views on necessity (which were inherited by Quine) in §§3.21–25.

Leibniz may have had the first clear conception of what we now call logical necessity. On the one hand, he had the unoriginal idea of possible worlds – worlds that it is possible for God to create. That notion was an intimate part of high scholasticism. On the other hand, he had the remarkable idea of what is provable from identities and definitions – that is, what is true in all possible worlds – and explains why it has to be true. It is no limitation on the powers of divinity, who chooses among possibilities.

A well-known logician once said to me that Leibniz had the idea of a completeness theorem: that truth in all possible worlds coincides with provability. As we now put it, a semantic idea, truth in all possible worlds, coincides with a syntactic idea, namely provability, or derivability in a finite number of steps.

Well, such completeness turns out to be remarkably non-obvious. Kurt Gödel proved the first great completeness theorem, the completeness of first-order logic, and thereby implanted the idea of completeness clearly in the mind of all future logicians. Gödel is often described as a platonist. Yes, but he was even more a leibnizian, believing that once we had adequate ideas (a far deeper understanding of sets than is at present available), we might have a complete set theory. That would fulfil what Gödel started, with his proof of the completeness of first-order logic, but it would take us no distance towards the ultimate, leibnizian completeness theorem, that truth in all possible worlds coincides with provability. Yet Leibniz also thought this hinged on getting the right ideas, the right concepts, not just of sets, but of everything.

21   Leibnizian proof

Some decades ago I had the gall to open a lecture (Hacking 1973) with the words: 'Leibniz knew what a proof is. Descartes did not.' I wish I were still an adolescent of the mind, and could allow myself to stroll today in the garden of unprotected aphorisms. Instead I have to say that Leibniz had a prescient grasp of the concept of proof we are now taught in elementary logic. A-proof-is-a-finite-sequence-of-sentences-each-of-which-is-either-an-axiom-or-follows-from-preceding-members-of-the-sequence-by-one-application-of-a-rule-of-inference. 'My claim for Leibniz [I continued] is only that he knew what a proof is. He was not even good at writing down proofs that are formally correct, for by nature he was hasty, in contrast to Descartes, who despised formalism and who is nearly always formally correct.' I would no longer say that Descartes despised formalism, for one can regard his algebrization of geometry as the introduction of a new formalism. Nor would I say now that he was nearly always formally correct, but the contrast between the two great philosopher–mathematicians, in respect of proof and truth, is plain enough. I shall not here give textual justification for these brash generalizations about the two philosophers; for some of that, see Hacking (1973).

Leibniz's account of proof was central to his amazing doctrine of truth. He had the implausible conviction that contingent truths have infinitely long proofs, 'or I know not what truth is'. Perhaps this bizarre idea can best be imagined in terms of Gerhard Gentzen's proof theory of the 1930s (Hacking 1974). We have absorbed only half of Leibniz on truth, namely the idea that necessary truths require only a finite number of steps. This is the core twentieth-century idea of a logically rigorous proof, though some would urge caution on the grounds that there may be necessary truths not provable in a finite number of steps.

Leibniz was well aware that the mechanical checking of such a proof, in order to be sure every step is in order, is beyond the patience and perhaps the ability of human beings. Luckily he invented the calculating machine. So too did Pascal. But Pascal's, I believe, was an interesting and potentially useful device in his own day. Leibniz's machine was a glimpse into the future, when, by the use of a Universal Characteristic, and a computer, one would be able to calculate or simulate the calculation of *everything*. He thought it would add immeasurably to the Art of Discovery, and he would have loved computer-generated proofs. For the fascinating road from Leibniz to Turing, see Davis (2000).

Experimental mathematics (§2.18) – not just proof-checking but also mathematical exploration by computer – can be seen as a lineal descendant of Leibniz's imagination. So can the simulation of the real world by computer, an absolutely standard practice in almost all the sciences, but one with which philosophers are only beginning to come to terms. Although I assert that the branching of the ways, between cartesian and leibnizian proof, appears in the seventeenth century, it was, like so many of Leibniz's speculations, not significant until fairly recently.

22 Voevodsky's extreme

Perhaps we are hourly becoming more leibnizian. Vladimir Voevodsky (b. 1966, Fields Medal 2002), worried in a lecture at the Princeton Institute for Advanced Study (Voevodsky 2010b) that mathematics as we know it, and as analysed in present-day Foundations of Mathematics, might be inconsistent. Contrary to the opinion of almost every other influential mathematician since, say, 1935, he thought it was more likely to be inconsistent. But that would, perhaps, be 'liberating'. For we could then use reasoning closer to ordinary 'intuitions' to develop ideas (all within an inconsistent framework). After that, we would develop proofs in some fragment of maths known to be sound.

Most philosophers and logicians have jeered at Wittgenstein's asking, what's so great about consistency? Could we not do perfectly good mathematics from an inconsistent basis, so long as we did not know it was inconsistent? Worse, those philosophers who have tried to make sense of what Wittgenstein wrote, tend, in my opinion, to trivialize it. Voevodsky is totally serious.

He won his medal in a branch of mathematics where proofs are very long, too long for one person, or even a team, confidently to check them – a phenomenon discussed below, §3.11. After the work that won the medal, he 'became convinced that the most interesting and important directions in current mathematics are the ones related to the transition into a new era which will be characterized by the widespread use of automated tools for proof construction and verification' (Voevodsky 2010a).

I owe this information to Michael Harris,* who has also allowed me to use some information from a work in progress. He reports a discussion in February 2011, at an informal gathering at the Institute for Advanced Study, Princeton. 'Voevodsky predicted it would soon be possible to design *proof-checkers* based on univalent foundations [a field that Voevodsky (2010a) is developing], that could effectively verify correctness of proofs written in the appropriate machine-readable language. In a few years, he added, journals will accept only articles accompanied by their machine-verifiable equivalents' (Harris, forthcoming, ch. 3).

Voevodsky's speculation is an extreme version of leibnizian proof. There is, of course, a lingering worry about regress. An author submits a paper with a proof or proof sketch, together with a programme for checking the proof, and a confirmation that, when run, the computer says, 'OK'. Who checks that the programme is sound? (See §33 below.)

23   Cartesian proof

Descartes did not have (or did not dwell on) Leibniz's sentential idea of proof at all. Descartes, as many scholars have observed, was not all that 'modern'. He had an older conception of proof that is still alive in the breast of almost every research mathematician.

In Cartesian methodology, in order to grasp the truth, you should get an entire proof in your mind all at once. He explains this, not in his mathematical writings, but in the lucid advice towards the end of *Meditation* I. Do not go quickly through my arguments, he urged, but master each one in its entirety, and hold it there before your mind. You must be able to run through a proof as a whole, and see it whole, before you properly grasp it. Reasoning is self-authenticating only if completely purified, completely stripped even of steps, so that you can grasp an entire proof as a whole. That was in effect Socrates' advice to Meno's slave. The lad discovered a fact, and a proof of the fact, thanks to Socrates' series of leading questions. But to fix the understanding, he must run through the exercise many times. 'These opinions have now just been stirred up like a dream; but if he were repeatedly asked these same questions, in various ways you know that in the end his knowledge about these things would be as accurate as anyone's' (*Meno* 85d).

24   Descartes and Wittgenstein on proof

It is tempting to co-opt the apt words used by Wittgenstein's translators, 'perspicuous' and 'surveyable', and say that Descartes wanted proof to be both. Here is Wittgenstein's key sentence of the late 1930s:

> Perspicuity (*Übersichtlichkeit*) is part of proof. If the process by means of which I get a result were not surveyable (*übersehbar*), I might indeed make a note that this number is what comes out – but what fact is this supposed to confirm for me? I don't know 'what is *supposed* to come out'. (*RFM*: 1, §154, 95)

I shall avoid those two tempting words, partly because I do not want to fall into the trap that Wittgenstein himself disparaged, of turning his catchwords into jargon. But in this case there is a further reason: it would take us

into difficult matters of interpretation – and here I mean not just hermeneutics but also translation from one language to another. As Baker and
Hacker (2005: 307, cf. 1980: 531) write, 'The notion of *Übersichtlichkeit* is
prominent in all Wittgenstein's later philosophy and is of paramount
importance', and 'it looms large in his philosophy of mathematics'. Before
they proceed to textual analysis, they devote several pages to the word and
its cognates, never mentioning mathematics, but usefully providing some
historical background. They also note that the published translations of
Wittgenstein vary from context to context, sometimes giving 'perspicuous'
and sometimes 'surveyable'. Hence they themselves opt to use the German
word, 'or some cognate of "survey" (including the archaic noun "surview")'.

If you are inclined to use Wittgenstein's words, you may find it useful to
observe that he *introduced* them, in connection with maths, in the quotation
above. Both *Übersichtlichkeit* and *übersehbar* are used in §54. Thereafter
he *quoted* those sentences, marked in quotation marks, and commented
upon the words. There is a sense (Quine's) in which he hardly every *used*
the words in connection with mathematics after their first usage; rather he
elucidated what he had meant.

Here is a roster of what I have in mind, with Wittgenstein's quotation
marks. He is talking about his own phrase from the late 1930s. I have
pedantically included snippets of the German original to observe that
even the same sentence has been rendered in English by different, often
equally apt, phrases – compare *übersehbar* in the third and fourth quotations. To repeat, these are *his* quotation marks. There are now many
discussions of Wittgenstein and perspicuity, but many authors quote
these sentences out of context and out of quotation marks, thereby masking
their role in Wittgenstein's thought.

'A mathematical proof must be perspicuous.' ['*Ein mathematischer Beweis
muß übersichtlich sein*'.] (*RFM*: III, §1, 143)

'A proof must be capable of being taken in' ['*Der Beweis muß übersehbar
sein*'] means [*meint*]: we must be prepared to use it as our guide-line
[*Richtschnur*] in judging. (*RFM*: III, §22b, 159)

'Proof must be capable of being taken in' ['*Der Beweis muß übersehbar sein*']
really means [*heißt eigentlich*] nothing but: a proof is not an experiment.
(*RFM*: III, §39c, 170)

'Proof must be surveyable': ['*Der Beweis muß übersehbar sein*'] this aims at
drawing our attention to the difference between the concepts of 'repeating a
proof', and 'repeating an experiment'. (*RFM*: III, §55c, 187)

When I wrote, 'proof must be perspicuous' ['*der Beweis muß übersichtlich sein*'] that meant: *causality* plays no part in the proof. Or again: a proof must be capable of being reproduced by copying. (*RFM*: IV, §41a, b, 146)

'A mathematical proof must be perspicuous' ['*Der mathematischer Beweis muß übersichtlich sein*'] – this is connected with the perspicuousness [*Übersichtlichkeit*] of that figure. (*RFM*: VII, §27f, 385)

On each occasion Wittgenstein provides his own commentary on his own words. Mühlhölzer (2010) provides 108 pages of commentary on the first twenty-two sections – sixteen pages – of *RFM* part III. The metaphors of Surveyability & Co. are wonderfully attractive and suggestive, but you may see why I prefer to evade the thickets of translation, interpretation, commentary, and hermeneutics.

25   The experience of cartesian proof: *caveat emptor*

In Chapter 3 I discuss why there is philosophy of mathematics at all. One answer is that a certain type of philosophical mind is deeply impressed by *experiencing* a Cartesian proof, of seeing why such-and-such *must* be true. This is quite unlike, for example, arithmetical computation, or in general following a rule by rote. I believe that Wittgenstein was from time to time obsessed by this experience, but I will not argue the case here. He insisted that proofs should be surveyable, in part because cartesian proofs got him so engrossed in philosophical issues about maths.

It might be wise to draw something like the opposite conclusion, partly because of changes in the sheer phenomena of mathematical proof that have arisen in the past half-century. Many proofs are so long that they can at best be leibnizian – a recent two-volume proof is mentioned in §3.11 below. 'Only a structure whose reproduction is an easy task is called a "proof"' (*RFM*: III, §1). Since the two volumes are in print, they are quite easily Xeroxed or scanned. But it is not that sense of reproduction that Wittgenstein had in mind.

We might venture this: whenever anybody wants to give an exemplary proof, they produce a lovely cartesian proof. That is just what I do, in my fourth epigraph, the proof of a theorem about not being able to dissect cubes to which I turn in a moment. But that is misleading because these ideal proofs are rare. Today's mathematical reality is largely leibnizian, not cartesian. It will become more so, if or when mathematics journals demand that every proof be accompanied by a proof-checking programme, as suggested at the end of §22 above.

So here is a paradox of which one must be aware. On the one hand, cartesian proofs are very often adduced as exemplars of what is puzzling about mathematics. At the same time, most mathematics is not cartesian, and may increasingly become leibnizian.

26 Grothendieck's cartesian vision: making it all obvious

Here is a quite different way to think about mathematics. Proof is important. More important are what are best called *proof ideas*: techniques of proving that rapidly generalize, perhaps by unexpected analogies, into many at first unrelated fields. There are mathematicians who conceive of success when they have finally transformed a field so that all the fundamental facts are 'obvious'. One strong candidate for true mathematical greatness is Alexander Grothendieck (b. 1928, Fields Medal 1966; stopped doing mathematics, in effect declining a Chair at the Collège de France and withdrawing to a cottage in the Pyrenees, 1970). By universal consent he

> is one of the most important mathematicians of the second half of the twentieth century, to whom we owe in particular a complete rebuilding of algebraic geometry. This systematic rebuilding permitted the solution of deep number-theoretic problems, among them the final step in the proof of the Weil Conjectures by Deligne, the proof of the Mordell Conjecture by Faltings, and the solution of Fermat's Last Problem by Wiles. (Scharlau 2008: 930).

(Pierre Deligne (b. 1944) won a Fields Medal, 1978, for his work on the Weil conjectures, and has just won the Abel Prize, 2013.)

Grothendieck was given to many rather apocalyptic statements. Here is one taken from the 2000+ pages of his *Récoltes et semailles* (roughly, 'Harvests and Seeds'; better, 'Reaping and Sowing'). In 2010 he forbade any of his work to be printed or distributed, so Springer stopped publication of a print edition. On page 368 of the cyclostyle version he wrote that '*mathematical creation*' is

> a vision that decants little by little over months and years, bringing to light the 'obvious' thing that no one had seen, taking form in an 'obvious' assertion of which no one had dreamed . . . and that the first one to come along can then prove, in five minutes, using techniques ready to hand.

Since the now-forbidden manuscript is not to hand, I have taken the quotation from Chapter IV of a new book by Michael Harris (forthcoming) mentioned already in §22. Incidentally, he too works extensively in the Langlands programme.

Yet we must backtrack. If the real aim of deep and fundamental mathematics is, as Grothendieck said, to create the obvious, then all the long, boring and non-memorable proofs are just so many crutches for the halt and lame, and they may be cured at the end of the mathematical road. We might say it is this 'obviousness' for which Descartes strove. Michael Harris takes Wittgenstein's idea that proofs should be perspicuous as a helpful way to understand Grothendieck's vision, but for the reasons just given, I abjure that.

Putting Grothendieck alongside Voevodsky, one invents a sort of parody *in extremis* of my contrast between cartesian and leibnizian proofs. It distorts the philosophy of both mathematicians, but it is nonetheless an instructive pairing. Which vision do you want? Grothendieck's extreme cartesianism? Or Voevodsky's extreme leibnizian idea (§22) that in the future all published proofs must be checkable on a computer? These are *not* incompatible. Unlike William Blake, most of us are unable to see fourfold, but I do believe we must accommodate both visions just suggested. We should strive to keep from Single vision: I refer to the last lines of a poem that Blake sent to a friend in 1802: 'May God us keep / From Single vision & Newton's sleep!' (Blake 1957: 818, lines 87–8).

27   Proofs *and* refutations

My second epigraph uses a few lines from *Proofs and Refutations* (1963–4, 1976) by Imre Lakatos (1922–74).* His book is often admired for its pedagogical wisdom – this is, how maths should be taught, not as abstract formal proofs, but as a process of discovery. The work was a continuation of the reflections of his great countryman, Georg Polya (1877–1985), of *How to Solve it* fame (1948; cf. 1954). In fact, Lakatos' book, in translation, had a very large print run in the Soviet Union, because its dialectic displayed the materialist rather than the bourgeois idealist view of mathematical reasoning. Good renegade communist that he was, Lakatos was amused that he was paid royalties for the labour involved, namely the number of words rather than a percentage of the actual sales. In the West, the book was greatly respected as damaging or destroying the myth of the certainty of mathematical proof. But it is seldom seen as the profound work of philosophy that it is.

It is the graphic story of how proof candidates, often widely accepted as decisive proofs, succumb to counter-examples. There is first a debate as to whether the example is just a freak, a monster, that should be excluded from consideration ('monster-barring'). That is an ad hoc, 'degenerating' strategy, to save the proof but learn nothing. The 'progressive' way is to modify the concepts, so that one has a deeper understanding of the proof idea, and is able not to limit its consequences (by excluding monsters) but, by expanding its domain, creating richer mathematics. All of this is brilliantly illustrated by Euler's 'theorem' about the lines, edges, and vertices of a polyhedron.

It is sometimes the case that in the course of such a dialectic, the connections between concepts become tighter and tighter, so that this is a road to the 'obviousness' at which Grothendieck aimed. And of course one of the things that Lakatos detested when he left Hungary for England in 1956 was the gross formalism then current in Western mathematical writing – an excess of Leibniz. *Proofs and Refutations* can be read as a manual on one of the ways in which Cartesian proofs come into being – by the clarification of ideas by resolving refutations.

In conversation, though not I think in print, Lakatos occasionally spoke of this as a process of *analytification*, of turning a 'synthetic' proposition into an 'analytic' one. When we get to that stage, a counter-example is no longer to be sought for – for the words are so used that they are bound to be true. One can give a sentential, a 'nominalist', cast to the idea I attribute to Lakatos. One is creating new sentences which end up as analytic – true in virtue of the words used. Before the culminations of a long process of proofs and refutations, no one has caused the words to be used in *that* way. Lakatos never wanted to settle on this idea, for it violates his first, falsificationist, principles.

Lakatos may well have been thinking of the same phenomenon that led Grothendieck to describe as the 'bringing to light the "obvious" thing that no one had seen, taking form in an "obvious" assertion of which no one had dreamed'. Grothendieck speaks of seeing what no one had seen before. Like 'stout Cortez when with eagle eyes / He star'd at the Pacific'. The obvious mathematical thing, that no one had seen, is, for Grothendieck, like the Pacific Ocean, out there. It is the recurring metaphor of exploration, which we find again and again in the work of mathematicians, and which reminds people of Plato.

To judge even by the sentence used in my epigraph, I don't think Lakatos had that idea at all. And although for a moment I dressed him up in 'nominalist' sententialist talk, that is only to misrepresent. His fundamental idea was that mathematicians construct what they discover. Their work becomes 'objective' as it becomes alienated from their productive activity. There was a great deal of Kant, Hegel, and some Marx in his way of thinking, and absolutely no 'platonism'. I shall not develop that theme here.

I have said in §1 that the experience of proof is one of the things that causes there to be a philosophy of mathematics. But that was too narrow. What counts is the lived experience of the entire mathematical activity of bringing a proof into existence, and that includes constant testing by counter-examples. That is an old meaning of the verb to prove: to test, as in the proof of the pudding is in the eating, and printer's proofs (alas, a lost techno-practice). And short of that, for those of us less able to do mathematics, there is at least the lived experience of working through a proof until light dawns.

One may compare Lakatos' idea of analytification to Wittgenstein's idea that in the course of a proof the very sense and use of the words is altered. Lakatos would have loathed the comparison, for he thought Wittgenstein was a very bad happening. But also, Lakatos' story of Euler is a story of how the boundaries of concepts were changed in historical fact, whereas Wittgenstein was speaking in the try-out mode, 'it is as if . . .', 'one might want to say here . . .'.

I hope it is obvious that this updated use of Leibniz's idea of analyticity (analytification) has nothing to do with the arid debates about the 'analytic-synthetic distinction' that followed Quine's brilliant denunciation of the account of logical necessity favoured by logical positivism.

28   On squaring squares and not cubing cubes

After these glimpses of work I do not fully understand, but which cannot be ignored, I return to the immediately intelligible. Many words can be used to characterise cartesian proofs. For example, they enable one to see not only that something is true, but why it is true. They give the feeling of understanding the fact proven. Others talk of the proof explaining the fact that is proved. Such words – *understand, why, explain* – are sound, but do little more than point to a satisfying phenomenon that is experienced, rather than one that can be well defined.

Everybody has their own favourite example of a proof, and usually, if not always, it is a cartesian one. Plato's example was constructing a square twice the area of a given square, a centrepiece of *Meno*. Hardy (1940) used two examples, in order to give his non-mathematical readers some idea of the experience of doing mathematics. One was Euclid's proof (*Elements* IX, 20) that there is no greatest prime. The second is 'Pythagoras' proof of the "irrationality" of $\sqrt{2}$'. Hardy (1940: 94, 109) uses these to good effect, for example in explaining why the second is much 'deeper' than the first.

My long-time favourite (used in Hacking 1962: 50) is the proof that I used as my fourth epigraph, the proof that a cube cannot be dissected into unequal cubes. I offer it as a cartesian proof in part because it will be new to most readers. It is important to have the experience of proof fresh in one's mind, and not to rehearse lovely old chestnuts. This does not mean that the cube theorem is unproblematic. It is certainly not formal. It employs *reductio ad absurdum*, which is not without its challengers. It is thoroughly pictorial, although it shows that the desired picture, a cube dissection, does not exist.

What is it to show that a construction cannot be found? What does finding a construction of a geometric figure with ruler and compass do to our conception of that shape? As a curious aside, we note that Wittgenstein (1979: 66) tortured his 1939 class in Foundations of Mathematics endlessly with that question – and Turing got rather annoyed both at the torture and at what Wittgenstein said. At issue was a student discovering how to construct a regular pentagon, by analogy with the construction of a regular heptakaidecagon – a 17-gon. In 1796, at the age of nineteen, Gauss had proved that it is possible, a beautiful example of the application of algebra to geometry – although the first actual geometrical construction was only made a few years later. I am sure that the example arose in Wittgenstein's 1939 class because in 1938 Hardy and Wright's *Introduction to the Theory of Numbers* (1938) had a new construction, much admired in Cambridge at the time.

The proof of the undissectability of a cube into unequal cubes is an unusual example of a thought experiment in mathematics, and of what Brown (2008: ch. 3) calls a picture-proof – although here it is an experiment that cannot work, a picture that cannot be pictured. (I am sure we should be cautious with the tendency to adduce picturesque proofs as cartesian, but the example does have its own interest.)

I took the proof from J. E. Littlewood's *Mathematician's Miscellany*. As is well known to mathematicians, Littlewood (1885–1977) was G. H. Hardy's (1877–1947) long-time collaborator at Trinity College, Cambridge. In a semi-popular essay, Littlewood (like Hardy in this respect) wanted to begin with 'mathematics with minimum "raw material"'. His words introducing the proof are worth following up:

> *Dissection of squares and cubes into squares and cubes, finite in number and all unequal.* The square dissection is possible in an infinity of distinct ways (the simplest is very complicated); a cube dissection is impossible. The surprising proof of the first result is highly technical. (See Brookes, Smith, Stone, and

Tutte, 1940.) The authors give the following elegant proof of the second. [Consult my fourth epigraph.]

Apparently around 1930 a Russian mathematician had conjectured that a square dissection was not only difficult but impossible – Lusin's conjecture. Wrong: in the period 1936–38, the Brookes, Smith, Stone, and Tutte named by Littlewood were undergraduates at Trinity College. They developed a hierarchy of dissections of squares. The impossibility of a cube dissection was an accompanying observation. (The first published example of a single square dissection, however, came from Göttingen (Sprague 1939).)

29 From dissecting squares to electrical networks

One origin of the dissection problem was a book of puzzles which was, a long time ago, well known to English schoolboys doing maths: Dudeny (1907). It has an imperfect squaring of a square, which might be a clue to a perfect squaring. Here is the puzzle of Lady Isabel's Casket.

> When young men sued for the hand of Lady Isabel, Sir Hugh promised his consent to the one who would tell him the dimensions of the top of the box from these facts alone: that there was a rectangular strip of gold, ten inches by ¼-inch; and the rest of the surface was exactly inlaid with pieces of wood, each piece being a perfect square, and no two pieces of the same size. How large was the box? (Dudeny 1907: 68)

More on puzzles in §2.27 below.

I shall allow myself a few anecdotes about the example, the first being a mere pun. As befits this chapter's title, 'A cartesian introduction', the four boys who solved the problem were often to use the pen-name *Blanche Descartes*, the 'Blanche' being constructed out of the initials of their first names. The 'B' in 'Blanche' was for Bill, namely William W. Tutte (1917–2002), the fourth-named author. His later major contributions to graph theory and other fields are described in the ten-page obituary for the American Mathematical Society (Hobbs and Oxley 2004). The 'C' was for Cedric, namely C. A. B. Smith, who became a distinguished statistician and geneticist.

For the romantically inclined, Tutte worked at the Bletchley Park code-breaking establishment during World War II that is famed for Alan Turing and the cracking of the Enigma code. Apparently Turing did not want Tutte in his group, so he became involved in the Lorentz code, used for communication among members of the German High Command. Tutte figured out how it worked thanks to a single error – a German operator did what was absolutely *verboten*, namely sent a long message twice, using the same machine, which changed the encoding. The two encodings of the same long message revealed on what principles the machine worked. Tutte inferred the logical structure of the machine, although no one had ever seen one. Tutte's solution was then used to build decoders called Colossus – they really were enormous calculating machines. (For how it was done and its consequences for the war, see www.codesandciphers.org.uk/lorenz/fish. htm and www.codesandciphers.org.uk/lorenz/colossus.htm.)

Less anecdotally, the solution to the squaring of the square problem (as it has become known) usefully illustrates how a mathematical problem can be attacked by using a seemingly unrelated piece of mathematics. The application of mathematics to mathematics – or is this the application of mathematics to physics? Or is it just the working out of precise analogies?

> It is, of course, not the puzzle but what they [the four undergraduates] made of it that is remarkable. To begin they translated the problem to the language of electrical networks. In the 1940 paper that later described their work are formulas for electrical network functions, not just those found earlier by Kirchoff, but new ones for transfer functions. This paper became a standard reference for electrical network practitioners. The question of squaring the square, in electrical terms, became a study of rotational symmetry of a part of the network, and how reflection of the symmetric part can alter its currents without affecting potentials on its boundary. The level at which they conceived the problem is remarkably deep. Their method did succeed in finding partitions of squares into smaller unequal squares. (www.squaring. net/history_theory/brooks_smith_stone_tutte.html)

This actually has it backwards, as is so often the case with history that uses hindsight. They did not get this method and then dissect. The 'Trinity Four', as they self-importantly called themselves, began in a typical exploratory way and produced a ranked hierarchy of square dissections by trial, error, luck, and some insight. They then saw that these correspond to electrical diagrams, and so were able to deepen and generalize their methods. Unlike with the cube theorem, one would never use their results, even in tidied-up form, to exemplify cartesian proof (Littlewood: 'very complicated'). The proof is not leibnizian either. Descartes and Leibniz prescribed ideal limits, and not much mathematics comes close to either ideal.

30 Intuition

To return to Descartes, he uses yet another word, the one we translate as 'intuition' when speaking of reason. His own statement in the *Rules for the*

Direction of the Mind is splendid. In the slightly non-standard translation by Anscombe and Geach:

> By *intuition* I mean, not the unwavering assurance of the senses, but a conception, formed by unclouded mental attention, so easy and distinct as to leave no room for doubt in regard to the thing we are understanding. It comes to the same thing if we say: It is an indubitable conception formed by an unclouded and attentive mind; one that originates solely from the light of reason, and is more certain even than deduction because it is simpler. (Descartes 1971: 155)

Deduction, he explains, means 'any necessary conclusion from other things known with certainty' (1971: 156). A necessary conclusion is *not* a conclusion that is necessarily true, or true of logical necessity. It is a conclusion that necessarily follows from the premises, so that if the premises are known with certainty, the conclusion will be known with certainty too. That's very different from *our* idea of logical necessity, which originates with Leibniz.

Stephen Gaukroger (1989) argues that the Cartesian doctrine of *intuitus* is a reaction against the late scholastic notion that reasoning is done by a mental faculty, one of several alongside the faculties of memory and imagination. To allow ourselves anachronism, Descartes was against the 'Modular Mind' that has become so entrenched in contemporary cognitive science. The Spanish scholars, whose texts were taught to Descartes in his boyhood at school in La Flèche, were the cognitive scientists of the day. At first sight this sounds wrong, for they held that logic was a normative theory about the right working of the reasoning module. In pedagogy, they brought logic and rhetoric together with this conception, and propounded rules for right reason, to which any reasoning must answer in order to be judged valid, and which students must use to verify their inferences. But beneath it all was something akin to an empirical theory about the mind, which divided into faculties or modules. In *Cartesian Linguistics* Noam Chomsky (1966) found Cartesian roots for his radical rewriting of linguistics. He could well have gone further back, to Spain, for his modular approach to the mind.

In *Rule III* Descartes says he is not trying to use the word 'intuition' in the sense of the schools, but rather in his own way derived from his understanding of Latin usage. And thereby hangs an ongoing problem. The concept of intuition is central to Kant, and there is an immense literature on *his* use of the idea. I am not well versed on Kantian scholarship involving the concept, and so will not pontificate on what he meant. In recent times it has become common for analytic philosophers to use the word 'intuition' to refer to their unreasoned hunches about what we would do or say in a bizarre situation. I deplore this practice. As the doyen of ordinary language philosophy, J. L. Austin, was apt to remark, we usually have no idea what we would say in a few words if, say, my cat delivered a philippic (his example): 'the only thing to do, and that can easily be done, is to set out the description of the facts at length. Ordinary language breaks down in extraordinary cases' (Austin 1961: 35). Finally, some cognitive psychologists now often use intuition to refer to any immediate inference in which there is no conscious reasoning. Others try to use cognitive science to explain what some mathematicians call intuitions, while others (e.g. Dehaene 1997) try to use the word as if there were a basic concept of 'intuitions' spanning Kant to ordinary hunches. These are among the reasons that I avoid the word. I shall say a little more about intuition, once again declining to use it for philosophical purposes, in §7.22.

To decline to use the word is not to despise it. Quite a few mathematicians find it a very useful word – to express the understanding that can be cultivated by thinking hard about a problem, and beginning to see how to solve it. Getting to know one's way around, and finally seeing matters clearly, compellingly – that is the very aim of mathematical activity. Many whom we shall encounter in this book would say that what makes their work worthwhile is not completing a formally correct proof, which is a total bore, but exploring the ideas preceding the final solution to a problem, seeing how they interact, and grasping how in the end they hang together. One finally has clear and distinct intuitions! As a teacher of research mathematics, one strives to create those intuitions in one's students. And that's what is important, and proof is merely a way of checking that one's intuitions were on the right track. I find this a very useful way of thinking about mathematical activities.

Why then do I decline to talk about intuition in this book? Partly because (as even Descartes reminds us, when trying to shed *his* predecessors' talk of *intuitus*) 'intuition' is like a pond in which many have waded, stirring up the sediment beneath, so that the water is too clouded to speak clearly. This is felt especially by a philosopher who hears his colleagues invoking what they call 'intuitions' for purposes that he thinks are better served by precise analysis.

31 Descartes *against* foundations?

Descartes rebelled against scholastic uses of the idea of intuition because he was a mathematician (and not a school or university teacher). He wrote as if he believed that *there are no normative standards to which reason can answer*. Often in reasoning we seek better, or earlier, reasons. But we must not demand a justification for reason itself. Reason is its own self-authenticating guarantor.

We just quoted the man: 'an indubitable conception formed by an unclouded and attentive mind; one that originates solely from the light of reason'. I suspect many working mathematicians would accept that as a description of what they want. It is what you get out of a cartesian proof – but not from a long, boring leibnizian exercise!

Readers of Descartes may protest that there is something else that authenticates reason, namely the good God who, in his bounty, would not trick us. I suggest that God is not there in the role of justifier. He is there in the role of *absence* of any possible justification. He is the background within which reason makes sense at all.

The role of the Cartesian God corresponds, abstractly speaking, to those elements in Wittgenstein's *Remarks on the Foundations of Mathematics* about what is prior to justification, 'bedrock', within which all discourse makes sense. I am willing to go further than texts could possibly justify, and say that this part of Cartesian theology plays a role analogous to Wittgenstein's 'natural history of man' (*RFM*: 1, §142, 92). Descartes, on that view, did not want a final justification of knowledge. If so, he was not the foundationalist that everyone takes him to be. But radical reinterpretation of Descartes is hardly our business here.

32 The two ideals of proof

Leibniz's theory of proof of a necessary proposition – a finite sequence of sentences (etc.) – became the ideal formalized by twentieth-century logicians. But we have another sense of proof, seeing as a whole, with clear conviction. That is the cartesian model, and it continues to serve us well. It is astonishing that we have not yet confessed to the duality of proof, cartesian and leibnizian. These are two ideals, which pull in different directions. We shall encounter cases where they are incompatible, e.g. the Monster (aka the Friendly Giant) and the classification of finite simple groups of §§3.8–12; they are what prompt Voevodsky's extreme leibnizian proposal of §22 above.

But that is only the beginning. Two ideals of proof do not cover the waterfront. Let's get real. It is only two distinct *ideals* that are encoded in the cartesian (get it all in your mind all at once) and leibnizian (sequence of sentences) notions of proof. These are crisp but artificial. There are many, many kinds of mathematical proof, few of which conform to either ideal. We also need to recognize that many arguments for the truth of mathematical propositions are rather compelling, but fall short of demonstrative proof.

One nice example is Euler's argument that the infinite sum of reciprocals of squares $(1 + 1/4 + 1/9 + 1/16 \ldots)$ is $\pi^2/6$. It is not a proof by any standards, yet it is almost wholly convincing. The argument is given in one of Georg Polya's wonderful books about heuristic reasoning (1954, §2.6). Both Mark Steiner (1973: 102–7) and Hilary Putnam (1975a: 67f) have used that as an example of *knowledge* of a mathematical fact based entirely on *plausible* argument.

Hacking (1962: 87) used an example from the same source (Polya 1954: 100) to the same effect, and quoted Polya's translation of Euler. Euler tests the numbers up to 20, and writes:

> I think these examples are sufficient to discourage anyone from imagining that it is by mere chance that my rule is in agreement with the truth.
>
> Yet somebody could still doubt whether the law of the numbers ... is precisely that one which I have indicated ... therefore, I will give some examples with larger numbers. (Polya 1954: 94)

He tests 101 and 301 and continues:

> The examples that I have just developed will undoubtedly dispel any qualms which we might have had about the truth of my formula. (1954: 95)

He goes on to say how the 'beautiful property of the numbers is so much more surprising' because it is completely unexpected. (No one could possibly imagine that it is 'analytic'!)

Of course I cite the example only because we now have a proof which, in Euler's words, makes 'intelligible the connection between the structure of the formula and the nature of the divisors'. We have learned by cruel experience that such confident inductions, in hands other than those of the incomparable Euler, often don't pan out. The eighteenth century was called, by its successor, an age of 'formalism' and 'inductivism'. Its methods were consumed by the new demands for rigour established by Cauchy (1789–1867) in his *Cours d'analyse* (1821), the first text of that title. His epsilon-delta formulation of the derivative will last as long as there is the calculus. (Curiously, the epsilon, the Greek 'e', was in the first instance short for 'erreur'.) Cauchy himself saw what he was doing as a return to Euclid (Grabiner 1981). But, as Putnam and Steiner insisted, the success of that 'revolution', or the fetish of rigour, does not mean that an argument that falls short of proof fails to be very good evidence. And as Kline (1980) observes, a great many of Cauchy's own publications did not meet his own demands for rigour.

33   Computer programmes: who checks whom?

The modern fast computer is manifestly the fruition of Leibniz's vision of proof. Yet every innovation creates its own problems. How do we know that we have an error-free programme? Donald MacKenzie (1993, 2001) shows how, as early as the 1980s, the tension between what I call the cartesian and leibnizian conceptions emerged in the heartland of commercial information technology. Some industrial consumers demanded proofs (cartesian ones) that the programmes were sound. This even resulted in lawsuits about whether something has been proven or not.

Roger Penrose has argued forcefully that computers are not (perfect) Turing machines. He draws on the now common observation that anything sufficiently complex is a quantum device, subject to quantum fluctuation. The simplest example is that a stray cosmic ray may at random turn a 0 into a 1 in the binary notation of a programme. The mathematical investigation of such issues demands serious philosophical reflection in the future. There is a substantial literature – amounting almost to a subdiscipline at the intersection of computer science and logic – on the reliability of computers. It is widely reported that almost half of the large fast computers in existence fall below IEEE standards, despite manufacturers' claims.

One can to some extent control for error using what may seem like a really klutzy idea, interval arithmetic. Computations are made using intervals rather than exact numbers; then one checks every exact number that one has computed to confirm that it lies within the interval computed. If it does not, there is a glitch somewhere. One might call this certainty controlled by uncertainty. Or double-entry book-keeping 2.0.

If you think (as Leibniz certainly did) that proof is a clear and distinct concept, then MacKenzie's history and Penrose's caution will be unsettling. They intended to unsettle! I am no longer troubled by the problems, real though they are. I see them as part of the continuing story of proof, from the time (as Kant put it – see §4.7 below) of 'Thales or some other' until today.

One thing that this past of ours is not, is orderly. We are uncertain how to carry on, with some insisting on something akin to cartesian standards, and others favouring the leibnizian idea. In my opinion, neither party is definitely right, or definitely wrong, because proof is an evolving concept, or group of concepts, and always has been.

Anyone who judges that proof is the essence of mathematics, or at least its gold standard, may now be inclined to step back a bit, and ask what makes mathematics mathematics.

CHAPTER 2

What makes mathematics mathematics?

1 We take it for granted

Philosophers, like most other people who think about it at all, tend to take 'mathematics' for granted. We seldom reflect on why we so readily recognize a problem, a conjecture, a fact, a proof idea, a piece of reasoning, a definition, or a sub-discipline, as mathematical. Some philosophers ask sophisticated questions about which parts of mathematics are constructive, or about set theory. Others debate 'platonism' versus 'nominalism', or, nowadays, versus 'naturalism' or 'structuralism'. But we seem to shy away from the naïve question of why so many diverse topics addressed by real-life mathematicians are immediately recognized as 'mathematics'. And what have those increasingly esoteric matters got to do with the common-or-garden mathematics of carpenters and shopkeepers; or, to move up a few social classes, of architects and stockbrokers?

Richard Courant published his classic *What is Mathematics?* in 1941. The Introduction ends with these words. 'For scholars and laymen alike it is not philosophy but active experience in mathematics itself that alone can answer the question: What is mathematics?' (Courant and Robbins 1996: n.p.). Courant was surely right: you learn what mathematics is by doing it.

In his rich essay prompted by being asked to reflect on beauty in mathematics, Robert Langlands (§1.15), after observing that mathematics 'is the work of men, so that it has many flaws and deficiencies' – and then that 'humans are of course only animals' – remarks:

> It is also very difficult even to understand what, in its higher reaches, mathematics is, and even more difficult to communicate this understanding, in part because it often comes in the form of intimations, a word that suggests that mathematics, and not only its basic concepts, exists independently of us. This is a notion that is hard to credit, but hard for a professional mathematician to do without. (Langlands 2010: 5)

That is the familiar dilemma recurring in debates between what is called 'platonism' on the one hand, or 'nominalism' and/or 'naturalism' on the other. It is very hard to comprehend how mathematics can exist independently of us, so we are all nominalists or naturalists at heart. But it is hard for a mathematician to do without this idea, so we are all platonists at heart. Yet to ask what makes mathematics mathematics is inevitably to foray into those perennial debates. Let us evade them as long as possible.

Obviously these very words, 'What makes mathematics mathematics?', can be used to ask very different questions. We might be asking an almost linguistic question: why do we call some things mathematics and not others? We might be engaged in metaphysics, asking what mathematics *really is*, often reduced to those various 'isms' just mentioned. But let us begin in as lowbrow a way as we can, by consulting a few dictionaries. Schoolchildren are forced to learn what is often called 'math' (USA) or 'maths' (UK), and so acquire an ability to use the noun 'mathematics'. What does it mean? Hence my initial resort to dictionaries.

2 Arsenic

Makers of desktop or larger dictionaries do their very best to capture, in a few lines, the meanings of words in their language. In the case of terms used in fields of expertise, they try to balance informed general knowledge and deference to recent science. Here, for example, is the definition of arsenic in *Webster's Third New International Dictionary of the English Language* (henceforth *Webster's*):

> **ar-se-nic** 1: 1: a trivalent and pentavalent metalloid element commonly metallic steel-gray, crystalline, and brittle but known also in other forms (as black amorphous and yellow crystalline forms), that occurs in the free state (as in tarnished granular or kidney-shaped masses having a sp. gr. of 5.73) and also combined in minerals (as aresenopyrite, orpiment, realgar, arsenolite) and in ores of other metals (as copper, gold) from which it is usu. separated as a by-product in the form of arsenic trioxide, and that is used in small amounts in alloys (as an alloy with lead for shot) and in the form of its compounds chiefly as poisons (as insecticides), in pharmaceutical preparations, and in glass – symbol *As* – see ELEMENT table.

There are two points to notice.

First, I could have used any standard dictionary in any widely used language, and the result would have been much the same. The only obvious difference is that most dictionaries put the atomic number 33 up front, while *Webster's* makes you look up the table of elements. My copy of *Collins* reads

'Symbol As; atomic no. 33; atomic wt.: 74.92; valency 3 or 5' – and continues with the relative density of grey arsenic, the melting point at 3 MN/m^2, and the temperature at which grey arsenic sublimates. That is all in a regular desktop dictionary.

A second point is that the format of these dictionary definitions is uncannily like the canonical form of definition commended in Hilary Putnam's (1975b) wonderful essay on the meaning of 'meaning'. Dictionaries do not give what some philosophers, with their characteristic contempt for ordinary people, dismiss as 'folk definitions': they combine general knowledge and well-established science in a single paragraph. Only readers of Agatha Christie will complain that *Webster's* does not highlight the poisonous qualities of arsenic that she and her successors made a part of common wisdom.

What's the point of introducing arsenic? First, to show how informative dictionaries are. Second, to show that the noun 'mathematics' is not at all like 'arsenic' or names of other substances so far as dictionaries are concerned. Different dictionaries give different accounts of what mathematics is. Here is an instructive sampling, with translations from major languages in which research mathematics is now published.

If you are prepared to take my word for it – that a sampling of dictionaries from around the mathematical globe shows many very different conceptions of mathematics – skip the next two pedantic sections. But word-lovers may enjoy the exercise.

3 Some dictionaries

math-e-ma-tics 1: a science that deals with the relationship and symbolism of numbers and magnitudes and that includes quantitative operations and the solution of quantitative problems – see FORMALISM 1d, INTUITIONISM 3, LOGICISM 2b. (*Webster's*)

math-e-ma-tics 1: the science of numbers and their operations, interrelations, combinations, generalizations, and abstracts, and of space configurations and their structure, measurement, transformations, and generalizations. (*Merriam-Webster's Collegiate Dictionary*, tenth edn, 1994)

mathematics. Originally, the collective name for geometry, arithmetic, and certain physical sciences (as astronomy and optics) involving geometrical reasoning. In modern use applied, (a) in a strict sense, to the abstract science which investigates deductively the conclusions implicit in the elementary conceptions of spatial and numerical relations, and which includes as its main divisions geometry, arithmetic, and algebra; and (b) in a wider sense, so as to include those branches of physical or other research which consist in the application of this abstract science to concrete data. When the word is used in its wider sense, the abstract science is distinguished as *pure mathematics*, and its concrete applications (e.g. in astronomy, various branches of physics, the theory of probabilities) as *applied* or *mixed mathematics*. (*The Oxford English Dictionary* CD-ROM)

math·e·mat·ics *n.* (*used with a sing. verb*). The study of the measurement, properties, and relationships of quantities and sets, using numbers and symbols. (*American Heritage Dictionary*)

matemática. Logical-deductive science in which the primary concepts are not defined (unity, conjunction, correspondence; point, straight line, plane) and propositions that are accepted without definition (axioms), from which are derived all of a theory for reasoning free of contradiction. (*Diccionario de la lengua española*. Espasa Calpe, 2005)

matemática. *f.* Deductive science that studies the properties of abstract entities, such as numbers, geometrical figures or symbols, and their relations. (In plural has same meaning as in singular.) ~**s aplicadas.** *f. pl.* Study of quantity considered in relation to physical phenomena. ~**s puras.** *f. pl.* Study of quantity considered in the abstract. (*Diccionario de la lengua española*, 22nd edn, 2001. Madrid: Real Academia Española)

au plur. Les mathématiques. Group (*ensemble*) of disciplines that proceed according to the deductive method and which study the properties of abstract objects such as numbers, geometrical figures and the relations that exist between them. (*Trésor de la langue française*)

LES MATHÉMATIQUES. Group (*ensemble*) of sciences, traditionally defined as sciences of quantity and order, which are characterized by their methods and by the fact that they give themselves their objects, which are abstract entities, postulated by their unique definitions (subject to the condition that they do not entail contradictions), and such that the set (*ensemble*) of their properties constitutes their essence. (*Le grand Robert de la langue française*, second edn, 2001)

رياضيات riyāziy[y]āt [Arabic: رياضیّات] (n.) The science of investigating numbers and the operations on them, relations between them, their combinations, generalizations, and abstractions, as well as figures and their structures and measurements and transformations and generalizations. **Modern mathematics** mathematics whose basis is set theory. **Applied mathematics** a branch of mathematics dealing with the use of mathematics in other branches of science such as physical sciences, biological sciences, and humanities; opp. pure mathematics. (Hasan Anvari *et al.* (eds.), *Farhang-e Bozorg-e Sokhan*, third impression. Tehran: Sokhan Publishers, 2007)

МАТЕМАТИКА, математики, no plural form, fem. A set of **disciplines that study quantity and spatial forms (arithmetic,** algebra, geometry, trigonometry, etc.). Pure m. Applied m. Higher m. (D. N. Ushakov (ed.), *Dictionary of Russian in Four Volumes,* 1935–40)

МАТЕМАТИКА. fem. A science, which studies quantity, quantitative relationships, and also spatial forms. Higher m. Applied m. (S. I. Ozhegov and N. Ju. Shvedova (eds.), AZ 1992)

數學[shu xue] Science dealing with quantity, shape, and the relationships between quantity and shape. Includes arithmetic, algebra, geometry, trigonometry, analytic geometry, differential calculus, and integration. Also known as "算學" (the study of counting). (*Online Mandarin Dictionary.* Taiwan: Ministry of Education, 1994)

數學 [shu xue] Science probing the space-form and quantity relationships in real life, including arithmetic, algebra, geometry, trigonometry, calculus, etc. (*Contemporary Chinese Dictionary,* Sichuan People's Publishing House, 1992)

数学 [sūgaku] (1) A field of study that investigates numerical quantity and space. It includes algebra, geometry, mathematical analysis (calculus and other branches) and their applications. (2) A field of study concerned with numbers, or what we know today as arithmetic. This usage was common from the time of the first publication of *Sūgaku Keimō* in China (1853), until the second decade of the Meiji period (1877–1886). (*Kojien,* sixth edn., Tokyo, 2008)

(*Sūgaku Keimō, An Introduction to Mathematics,* was a textbook of European algebra written in Chinese by Alexander Wylie (1815–87), who had a remarkable knowledge of early Chinese mathematics, as well as that of contemporary Europe; see Wylie (1853).)*

Ma| thema| tik, die. Science, theory of numbers, shapes, sets, their abstractions, and of the possible relations or connections between them. *applied M.* (Branch of mathematics which is concerned with numerical applications.) *pure M.* (Mathematics with no view as to its applications, and concerning itself only with mathematical structures.) (*Duden: der Grosse Wörterbuch der Deutschen Sprache,* 1994)

Für *Mathematik* gibt es keine allgemein anerkannte Definition. (German *Wikipedia,* accessed October 2010)

4 What the dictionaries suggest

After reading through this list, we may agree with the last statement, that there is no generally recognized definition of 'mathematics'. German *Wikipedia,* accessed May 2012, is a little more cautious: 'Mathematics is the science which originates with the investigations of shapes and calculations with numbers. There is no generally recognized definition of *mathematics*.' It continues by saying that nowadays one usually says it is the science that creates abstract structure by means of logical definitions, and investigates their properties and patterns by means of *Logic*. (This has been augmented a bit since then; happily there is no final wiki, ever.)

We think of mathematics as universal and transnational, but it is almost as if different linguistic traditions have different conceptions of mathematics.

On mathematics, *Webster's Third International*, the canonical American authority, is odd indeed, and seems never to have heard of geometry, while at the same time referring us to the mathematical/philosophical/ foundational debates of the first part of the twentieth century. That is perhaps unsurprising, for although it has been updated a little, the *Third* is a work of 1961; a *Fourth* is in preparation.

The desktop version, *Merriam-Webster's Collegiate*, the standard authority preferred by the *Chicago Manual of Style* (and not to be confused with the much inferior rival products of other companies that use 'Webster' in their title as a generic label, like Kleenex) usually abridges the big dictionary, but in this case much improves on it.

History. The *OED* reminds us that the word 'mathematics' has evolved over time.

One science or several? French dictionaries and one Russian one treat mathematics not as one discipline or science but as a collection of disciplines.

Abstract entities. The *OED* speaks of an abstract science, but only the French and Spanish dictionaries speak of abstract objects or entities. *Robert* speaks of these objects as having an essence given by their definitions.

Pure and applied. The *OED* and several other dictionaries distinguish pure from applied mathematics. One Chinese dictionary, reflecting dialectical materialism, insists that mathematics has to do with real life (pure mathematics would be filed away as bourgeois idealism).

Axioms. Only the Spanish definition mentions axioms. That is the most technical definition found in our list, but taken literally it would exclude arithmetic, and that for two reasons. First, arithmetic was not axiomatized until the time of Peano – say, 1900. So, unlike geometry, axiomatized at the time of Euclid, the theory of numbers would not count as mathematics until Peano. Second, arithmetic has no complete axiomatization, so only axiomatized parts of arithmetic would, under this definition, count as maths.

Structure. German, interestingly, characterizes pure mathematics as concerned with structure.

Deduction. The *OED*, both Spanish, and one French dictionary characterize mathematics partly in terms of deduction.

None of the dictionaries (except maybe *Webster's Third*!) is wrong; together they illustrate that mathematics has quite a few faces.

5   A Japanese conversation

Let us continue in the same vein for a moment, with the everyday world for which dictionaries cater, and not turn, yet, to specialist knowledge.

The novelist Haruki Murakami creates a universe of many worlds, mostly underground, and tinged with this or that bit of absurdity or magic, but all within a framework of ordinary banality. It is the sheer banality – of which he is a consummate master – that interests us here.

IQ84 (Murakami 2011) features Tengo, an aspiring novelist who teaches mathematics at a crammer intended to help students get into university. Part of the plot develops from his encounter with a very odd seventeen-year-old, Fuka-Eri, who has written a strange short story. On the banal side, they talk about mathematics, for Fuka-Eri has come to a couple of Tengo's classes. She always speaks in a completely flat voice, making no distinction between questions and statements, and uttering just one sentence at a time. One conversation between Fuka-Eri and Tengo runs like this:

> *Fuka-Eri.* You like math.
> *Tengo.* I do like math. I've always liked it, and still like it.
> What about it?
> What do I like about math? Hmm. When I've got figures in front of me, it relaxes me. Kind of like, everything fits where it belongs.
> The calculus part was good.
> You mean my lecture? [Fuka-Eri nods]
> Do you like math?
> [She gave her head a quick shake. She did not like math.]
> But the part about calculus was good?
> You talked about it as if you cared. (Murakami 2011: 45)

This is a perfectly ordinary conversation (albeit embedded in a larger framework of something strange). That is one of the ways that people talk about mathematics. Now Fuka-Eri's 'math' already suggests school. (I have not checked the original Japanese.) Part of what makes mathematics mathematics is that it is called that in school or college. In order not to scare children, schools change the name to make it sound more attractive, less scary, and less off-putting. But it stays the same. All the children know it is still maths. What do they recognize?

Maybe just something they find hard. Most people resemble Fuka-Eri much more than Tengo. He is no great shakes as a mathematician, but he does like mathematics. Indeed, most people shy away from mathematics, and many actively hate it. They do not enjoy the experience of mathematics, do not grasp proofs, and do not like playing around with structures – and that despite the popularity of Sudoku.

Nevertheless, it is as if there is *something* they respond to with dislike. So what is this thing called mathematics?

6   A sullen anti-mathematical protest

Yves Gingras (2001) draws attention to a Luddite protest in the history of the relation between mathematics and physics. In the seventeenth and eighteenth centuries, that fragment of what we call physics that was then practised fell under natural philosophy. In France there was *la physique*, studied by *physiciens*. In contrast, the English words 'physics', and especially 'physicist', were late coinages. But English and French sciences strongly overlapped and Gingras calls them by the modern name, physics. His paper has the informative title, 'What did mathematics do to physics?' As Husserl put it, Galileo mathematized the world, but in making mathematical analysis the way to understand how the physical world works, he also began to change physics. Well into the eighteenth century there was a wealth of natural philosophers fascinated by physical phenomena. But as Newton advanced on Galileo and Descartes, it became increasingly necessary to be a mathematician in order to be a physicist. And many physicists or natural philosophers hated that. Gingras quotes numerous protests.

Of course one side of settled human nature resists having to learn new things. (I often hate it when my word-processing programmes are upgraded for benefits of no interest to me.) But in addition these people feared what the mathematicians were doing to their beloved science, for they no longer had any status within their domain of interest. And they detested the new maths. To repeat the ending of the previous section, it is as if there were *something* they responded to with an emotion stronger than dislike. Is it just a bundle of techniques for which most human beings have no talent?

7   A miscellany

The impossibility of cubing a cube was used as my fourth epigraph and discussed in §1.28. I took it from Littlewood's wonderful little book of examples and anecdotes, *A Mathematician's Miscellany* (1953). His very title draws attention to the diversity of mathematical activity, so I co-opt it here. Let us survey the miscellany that is mathematics.

The arithmetic that all of us learned when we were children is very different from the proof of Pythagoras' theorem that many of us learned as adolescents. When we began to read Plato, we saw in the *Meno* how to construct a square double the size of a given square, and realized that the argument is connected to 'Pythagoras'. But that is totally unlike the rote skill of doubling a small integer at sight, or a large one by pencil.

Both types of example are unlike the idea that Fermat had when he wrote down what came to be called his last theorem. We nevertheless seem immediately to understand his question about the integers. The situation is very different from the proof ideas that lie behind Andrew Wiles' discovery of a way to prove the theorem. Few of us have mastered even a sketch of that argument. Is it 'the same sort of thing' as the familiar proof that there is no greatest prime? I am not at all sure.

Recent journalism has rightly brought Fermat's theorem to fairly general knowledge. One adage to be followed in the present series of thoughts is Wittgenstein's: don't have too slender a diet of examples on your plate. So let's add to our roster another fact that Fermat wrote down. Every prime number of the form $4n + 1$ is the sum of exactly two square numbers. (Thus $29 = (4 \times 7) + 1$, and is also the sum $25 + 4 = 5^2 + 2^2$.)

The incomparable Euler proved that theorem, which I personally find startling. This seemingly meaningless fact is true of *all* primes of that form. It is a perfect example of a beautiful and thoroughly unobvious result in number theory. How different it feels from Kant's boring paradigm, $5 + 7 = 12$. Whereas Kant's proposition is something one learns (or used to learn) by rote at an early age, Fermat's looks like a telling example of a 'synthetic' proposition. How could 'sum of two squares' be contained in, or be part of the analysis of, the concept 'prime number equal to $4n + 1$'?

Well, that is exactly what Frege set out to explain. By defining arithmetical concepts and using a pure logic that he himself constructed, he could derive such truths. Hence (he inferred) Kant was wrong: the proposition about primes of that form is analytic – in exactly the sense intended by Leibniz. Explicating (rather than reading) Leibniz only a little, an analytic truth is one derivable by logic from identities, or at any rate from definitions. Frege's *Foundations of Arithmetic* (1884) remains to this day the most compelling example of philosophical analysis, ever. That is why J. L. Austin, the doyen of Oxford 'ordinary language' philosophy, chose to translate it – impeccably.* When I say that Frege's analysis is compelling, I do not mean that it is uniquely right, or even that it is 'right', but rather that it had the power to change the landscape of numbers forever.

Yet, as Gödel taught us, no extension of Frege's technique could ever capture the whole of arithmetic, even. *Is* there a 'whole of arithmetic' to capture?

In passing we may notice that $5 + 7 = 12$ is an awfully useful fact. Indeed what interested Kant was not this sum as an example of what we now call 'pure mathematics', but as something readily used – 'applied' – in daily life. Five eggs in the fridge and seven on the counter: twelve in all.

It is almost as if the two facts about numbers – Kant's and Fermat's – inhabit different worlds. Could Fermat's fact have *any* use outside pure theory of numbers itself? Yes: I can use it to show off my erudition or as a running example of a non-obvious fact, but (to mimic Quine) that is not *using* the fact but *mentioning* it; many another fact would serve as well.

Now for a completely different scene. The mathematics of theoretical physics will seem a different type of thing from arithmetic or Euclidean geometry, but we should not restrict ourselves to theory. Papers in experimental physics abound in mathematical reasoning. Martin Krieger (1987, 1992) speaks of 'the physicist's toolkit', a surprising amount of which is a collection of rather old mathematical tools in modern garb – including Lagrangians (Joseph Louis Lagrange 1736–1813), Hamiltonians (William Rowan Hamilton 1805–65), and Fourier transforms (Joseph Fourier 1768–1830). (This does not mean that it is easy for today's mathematical physicist to read Fourier or even Hamilton, let alone Newton or Leibniz, inventors of the quintessential tool, the differential and integral calculus.) The mathematics in the physicist's toolbox – and the way it is used – looks very different from that of the geometer or the number theorist.

Something entirely new has been added to the tools of the physicist, indeed of all scientists and quite a few humanists. In the sciences we have powerful and increasingly fast computational techniques to make approximate solutions to complex equations that cannot be solved exactly. They enable practitioners to construct simulations that establish intimate relations between theory and experiment. Today, much – maybe most – experimental work in physics and chemistry is run alongside, and often replaced by, simulations. Is the simulation of nature by powerful computers applied mathematics, in the same way that modelling nature using Lagrangians or Hamiltonians is called applied mathematics?

Physics employs sophisticated mathematical models of physical situations. Economists also construct complicated models. They run computer simulations of gigantic structures they call 'the economy' to try to figure out what will happen next or in ten years' time. The economists are as incapable of understanding the reasoning of the physicist as most physicists are of making sense of modern econometrics. Yet they are both using what we call mathematics, and the skills are to some extent transferable. Witness the post-Cold War exodus of high-energy physics PhDs to Goldman-Sachs (etc.) a few years before the near collapse of the global banking system.

James Clerk Maxwell had a theorem about the necessary and sufficient conditions for the rigidity of a three-dimensional structure. It appeared in textbooks for many years. He was mistaken. The 'bucky domes' devised by Buckminster Fuller provide a counter-example (§5.25). What parts of this story are mathematics?

Are programmers writing hundreds of metres of code doing mathematics? We need the programmers to design the programmes on which we solve, by simulation and approximation, the problems in physics or economics. What part is mathematics and what part not? What about cryptography? (To avoid misunderstanding, I ask such questions in the rhetorical mode. In §27 below, I note a disagreement about whether chess problems are mathematics, but don't want to suggest there is a uniquely right answer. The maxim, 'say what you choose, so long as you understand what you mean', fits here.)

But to return to our miscellany, arithmetic for commerce seems very different from the theory of numbers. Geometry for carpentry is unlike the proof, to be found in Euclid, Book XIII, that there are exactly five Platonic solids (regular polyhedra).

Why, to return to the question posed in the last chapter, do we take for granted that arithmetic and geometry are both part of 'the same thing', namely mathematics?

What makes mathematics mathematics?

8　An institutional answer

As befits the start of any serious philosophical thought, my initial question aimed at naïvety. It is not technical. It could be rephrased in a way that loads the answer in one direction: what enables us to recognize as mathematical what we call mathematics? A very important answer is institutional. Mathematics is a discipline defined in institutions of higher learning by departments and even faculties. The three Rs were once obligatory: Reading, Writing, and Arithmetic. Is the third maths? We have all been subjected to the mathematical regime, even if only in primary school. Most of us hated it, a few of us loved it, and a small number became real persons only when they found themselves at home there.

Why are we leery about calling programming part of mathematics? Simple answer: because pedagogy created departments of computer science. Why don't we call chess mathematics? Because it is not taught in school or college. But one might reverse the question. Why should we even think that chess might be filed as a branch of maths? Isn't it because it has some characteristics typical of mathematical reasoning? That seems to take us back to our naïve question, which turns into: what are those characteristics? Likewise, although the organization of mathematics teaching is the upshot of a long and highly contingent history, is it not a history that focuses on those very characteristics?

9   A neuro-historical answer

There was once a vogue for writing psycho-histories of prominent persons, usually with a heavy dollop of Sigmund Freud. Recently this has been replaced by the diagnosis of autism, Asperger Syndrome, or even Attention Deficit Hyperactivity Disorder, as an 'explanation' of why some prominent people became prominent in the way they did. Since these disorders are now thought to be neurological and even heritable, we might call this the genre of neuro-history. For lush examples of this type of biography, see a series of books published by the flamboyant Irish psychiatrist Michael Fitzgerald, 2004–9. More seriously, there is the work of Simon Baron-Cohen and his collaborators (1997, 2002, 2003), suggesting that a range of trades, including engineering and mathematics, attract people with autistic traits. This is popularized in fiction by the new dogma that computer nerds are autistic.

The aspects of autism once grouped under Asperger Syndrome include social diffidence, difficulties in understanding what other people (or even themselves) are feeling, single-mindedness, and a fascination with detail and with patterns. Such individuals are easily upset by change, particularly in spatial arrangements or temporal sequences. Most readers familiar with that extensive literature, be it fact or fiction, will readily recognize autistic tendencies in their own behaviour. (The diagnosis of Asperger's has been withdrawn from the *Diagnostic and Statistical Manual of Mental Disorders*, but 'aspie' may well remain a common noun in English.)

Suppose that mathematicians tend to have such personality traits. It is an easy step to propose that their brains are especially sensitive to invariance and symmetry, two of the cardinal notions of modern mathematics. It has long been taught, by otherwise quite different schools of thought, that mathematics is the study of structure and order, a study with peculiar appeal to people with autistic tendencies.

This picture of individual idiosyncrasy fits nicely with the institutional story. Mathematics is what is done by institutions dedicated to the subject. These are populated by individuals who find homes where their peculiarities are the norm, which in turn cements the organization of the institutions.

These remarks conclude our discussion of neuro-history (thank goodness!), but they could not be altogether omitted.

10 The Peirces, father and son

Benjamin Peirce (1809–80) professed mathematics at Harvard for some fifty years, and was a seminal figure in the establishment of mathematics and the sciences in the United States. His son, Charles Sanders Peirce (1839–1914), founded pragmatism, although he was sufficiently displeased by later pragmatists that he renamed it 'pragmaticism', a name too ugly to steal. He was a bit like Leibniz, with brilliant ideas about almost everything, and yet with not a great deal of work brought to a satisfactory termination. (My own account of him as scientist, and as someone who really understood probability, is to be found in Hacking 1990: ch. 17).

In 1870, the father delivered a paper to the National Academy of Sciences in Washington, which asserted in its first sentence: 'Mathematics is the science which draws necessary conclusions' (B. Peirce 1881: 97). When published posthumously, its editors observed that the 'work may almost be entitled to take rank as the *Principia* of the philosophical study of the laws of algebraic operation' (1881). It may be tempting to read Peirce's definition as asserting that mathematical theorems are logically necessary propositions. That is not what Benjamin Peirce meant.

As already observed in connection with Descartes (§1.30), a necessary conclusion is one that necessarily follows from premises, not one that is true of logical necessity. In more recent language, he was saying that mathematics is the science of valid inference, of the science of logical consequence. Thus, anachronistically, he might be said to define mathematics semantically, while Russell, as quoted in §11 below, defined mathematics syntactically. On the relation between logic (as he understood it in 1870) and mathematics, Benjamin Peirce stated that: 'Even the rules of logic, by which it is tightly bound, could not be deduced without its aid' (1881).

In 1902 his son examined the 'essence' of mathematics (*The Simplest Mathematics*; CP 4: 227–323, 1902) – and began by recalling his father's definition. Charles Sanders Peirce retained the Kantian view of the relationship between logic and mathematics: 'It does not seem to me that mathematics depends in any way upon logic' (*CP* 4: 228). Both the father, a philosophical mathematician, and the son, a mathematically minded philosopher, have many interesting observations about the nature of mathematics, but with regard to our question, 'What makes mathematics mathematics?' they are but a way-station, telling us very little except what now seem to be platitudes or oddities.

11   A programmatic answer: logicism

Institutional answers invite histories of teaching, initiation, and authority as well as contemporary sociology. But one can hear the question, 'What makes mathematics mathematics?' in a very different way again. One wants mathematics itself to answer the question. I don't mean, let's ask mathematicians. I mean detailed mathematical answers, or programmes, which, if they succeeded, would appear to provide definitive answers. One familiar example is logicism. In fact the most daring answer to the ur-question occurs on the first page of Russell's *Principles of Mathematics* (1903):

> Pure mathematics is the class of all propositions of the form '*p* implies *q*', where *p* and *q* are propositions containing one or more variables, the same in the two propositions, and neither *p* nor *q* contains any constants except logical constants. And logical constants are all notions definable in terms of the following: Implication, the relation of a term to a class of which it is a member, the notion of *such that*, the notion of relation, and such further notions as may be involved in the general notion of propositions of the above form. In addition to these, mathematics *uses* a notion which is not a constituent of the propositions which it considers, namely the notion of truth.

One could argue that the entire logicist programme is a footnote to the paragraph just quoted. If the amazing three volumes of Whitehead and Russell's *Principia Mathematica* (*PM*) (1910–13) had wholly succeeded, our seemingly naïve question would have a direct answer. Something is mathematics if it is logic!

Logicism had far deeper aims than answering our question, but *PM* would in passing have served up an answer. Yet in a sense *PM* took mathematics for granted. It is as if Russell had said, we are well able to recognize mathematics, and in that sense we know perfectly well what it *is*. The task is to establish an informative *analysis* of what mathematics is, just as Frege gave an analysis of the concept of number. At one time Russell, ever in need of certainty, thought establishing that mathematics is logic would provide a foundation for mathematics, in the sense that it would underwrite its certainty. He meant a 'foundation' in the literal sense, not in the sense of the discipline now called Foundations of Mathematics. He meant a secure basis that would support and guarantee the truth of all mathematical theorems. Regardless of that possibly misguided aspiration, *PM* would have been a precise sophisticated answer to our question, 'What makes mathematics mathematics?'

Or would there have been a regress? Quine above all others posed the question, 'What makes logic logic?' Thanks to Frege and *PM* he could sharpen the question, agreeing with the paragraph just quoted, but going on to ask, 'What is a logical constant?' Russell's little list of constants has problems, especially – as Russell was the first to demonstrate – with the notion of set membership: 'the relation of a term to a class of which it is a member'. (For an anthology addressed to the question of what logic is, see Gabbay (1994) which begins (pp. 1–34) with my own answer (Hacking 1979). For a full survey with references going back to the schoolmen and on to the present, see MacFarlane (2009).) Without knowing what makes logic logic, *PM* could hardly complete the answer to what makes maths maths.

This observation is completely independent of the mathematical difficulties that arose in connection with that wonderful monument that is *Principia Mathematica*.

12   A second programmatic answer: Bourbaki

A second example of a mathematical answer to 'What makes mathematics mathematics?' came a generation after *Principia Mathematica*. A handful of talented young mathematicians, at the time teaching in the French provinces, met together with the intention of writing the 'Ultimate Mathematical Textbook' (Corry 2009). They named themselves Nicolas Bourbaki (Mashaal 2006). Their first volume, a distinctive treatment of set theory according to their own rather unusual standards of rigour, appeared as *Éléments de mathématique* (Bourbaki 1939).

When they reconvened at the end of World War II, they were increasingly convinced that mathematics was the study of structures. They took it upon themselves to present the whole of mathematics as a nested or otherwise organized treatment of structures. Successive volumes (in 'structural' as opposed to chronological order of printing) went from ii: *Algebra*, iii: *Topology*, through to viii: *Lie Theory*, and ix: *Spectral Algebra*. (The work on algebra is still widely cited, while the initial volume, on set theory, is not.)

As with Russell, they 'knew' what counted as mathematics: they could tell mathematics when they saw it. But they did not include the branch of mathematics called probability theory. They probably thought it was implicitly covered by VI: Integration, namely what others call measure theory. Probability theorists, including distinguished French ones, did not agree!

Just as Russell held that mathematics *is* logic, Bourbaki thought mathematics *is* the study of structure, which, as Corry explains, is a more malleable concept than they had at first believed. Corry's essay (2009) is invaluable secondary reading, for it points to the centrality of *structure* to Bourbaki, right from the start.

Here we must pause for clarification. 'Structuralism' has a number of meanings. The one that comes first to mind for the general reader is likely to be the French intellectual movement deriving from Ferdinand de Saussure's structural linguistics of the early twentieth century – and especially its flourishing in the 1960s, sparked by the anthropological reflections of Claude Lévi-Strauss, which created waves in sociology, literary criticism, classical studies, psychoanalysis, and much else.

The doctrines of Bourbaki do intersect with the ideas of Lévi-Strauss – they were part of the same intellectual climate in France directly after World War II. One of the central figures of the Bourbaki group was André Weil (recall §1.14), who wrote an appendix for Part 1 of the 1949 classic, *Elementary Structure of Kinship* – on the 'Algebra of certain types of marriage laws' (Lévi-Strauss 1969: 221–30). Lévi-Strauss (1954) wrote enthusiastically about such matters. For a helpful survey, 'Mathematical metaphors in the work of Lévi-Strauss', see Almeida and Barbosa (1990).

Structuralism of that sort will *not* come to mind for contemporary philosophers of mathematics. They will think of a movement often said to begin with a famous paper by Paul Benacerraf (1965), although often claiming roots in Dedekind's well-known essay (1888). Various kinds of doctrine have emerged, as discussed in Chapter 7B, but all take for granted standard 'denotational' semantics, and profess to engage in metaphysics and *ontology*, alien to Bourbaki. One of Bourbaki's heirs, André Lichnerowicz, characterizes his attitude as 'radically *non-ontological*' (Connes, Lichnerowicz, and Schützenberger 2000: 25, my italics).

Reck (2003) has a useful classification of various philosophical structuralisms available today (cf. Reck and Price 2000, and Reck 2011). Reck contrasts these philosophical structuralisms with what he calls 'methodological structuralism' as found in Dedekind. That 'has primarily to do with mathematical *method*, rather than with semantic and metaphysical issues as the others [structuralisms of analytic philosophers] do. Thus it is really in a separate category, or of a different kind' (Reck 2003: 371). The same may be said of Bourbaki: it is *in a separate category, or of a different kind* from those canvassed by recent analytic philosophers. We shall return to these issues in §7.11.

Both logicism and Bourbaki represented a totalizing moment in intellectual life. In the terminology of Lyotard (1984), they presented grand narratives (his original labelling was meta-narrative, which is not apt here). Russell and Bourbaki were epitomes of the 'Modern', as that term has often been used after Lyotard. Each left a monumental text that will never be imitated. About the only thing that Lyotard said or implied, specifically about mathematics, and with which one can sensibly agree, is that after Gödel no one will ever attempt an 'Ultimate Mathematical Textbook' or a *Principia Mathematica* again.*

Bourbaki and Russell did not ask the naïve question of what makes mathematics mathematics. Mathematics was a given; their task was to analyse it, and in that way to state what it *is*, in an explanatory way. I am not being dismissive. I was brought up in logicism and have learned to love Bourbaki. I say only that neither addressed the most naïve question of all.

13   Only Wittgenstein seems to have been troubled

Wittgenstein seems to have been the first notable philosopher ever to emphasize the differences between the miscellaneous activities that we file away as mathematics. Chapter 1 has already quoted his remark: 'I should like to say: mathematics is a MOTLEY of techniques of proof. – And upon this is based its manifold applicability and its importance' (*RFM*: III, §46, 176). The appended thought is hard to grasp. I find it much easier to see that maths is a motley of techniques of proof than to understand why its importance and manifold applicability depend on that.

Where the translators use a single capitalized MOTLEY, Wittgenstein's German has a capitalized adjective followed by an italicized noun: 'ein BUNTES *Gemisch* von Beweistechniken'. Wittgenstein is not given to such orthographic hyperbole; perhaps this is the only occasion in his published texts that he used it. So let us attend to it with care.

In Luther's Bible, *bunt* is the word for Jacob's coat of many colours, and the word in general means parti-coloured, and, by metaphor, miscellaneous. In current German it is a rather down-putting adjective, which is *not* to say that Wittgenstein was putting down the motley of techniques of proof.

Quite the contrary: the importance and 'manifold applicability' of mathematics, depends, he went on to say, on this motley.

'Motley' is an apt translation of Wittgenstein's double-barrelled phrase 'buntes Gemisch' because it implies a disorderly variety within a group. To compare: the German noun 'Treiben' denotes bustling activity; 'ein buntes Treiben' is emphatic, meaning a real hustle and bustle with all sorts of different things going on (more likely involving folks in the lower orders of society). Likewise, 'ein buntes Gemisch' is not just a mixture, but rather a mixture of all sorts of different kinds of things. When the noun is italicized in emphasis, and the adjective is printed in capital letters, BUNTES *Gemisch*, Wow!

Felix Mühlhölzer (2006: 66, n. 15) argues that 'motley' does not well represent Wittgenstein's usage. He prefers to translate BUNTES *Gemisch* as 'colourful mix'. Somehow I can't see someone whose prose is not given to orthographic hyperbole writing COLOURFUL *mix*. I will not argue, but the word 'motley' certainly captures, and perhaps exaggerates, what I have called a miscellany. But there is another aspect of 'buntes' = 'colourful' that 'motley' misses. 'Colourful' is a word that usually has positive connotations – colourful as opposed to drab. Mühlhölzer is right to bring out this aspect of 'buntes' that 'motley' misses. I shall nevertheless stick to Anscombe's original translation, but keep the *colourful* in the back of your mind.

§46 says only that techniques of proof form a motley. A few pages (but only two 'remarks') later, Wittgenstein said, with less emphasis, that he wanted 'to give an account of the motley of mathematics' (III, §48, 182). That sounds like a very general remark about maths.

The 'motley' metaphor reminds us of Wittgenstein's well-known family resemblances, but to speak of the motley (or even colourful mix) of so-and-so is far more emphatic than saying that instances of so-and-so form a family. This is not to deny that there are family resemblances among the miscellaneous examples of mathematics that I have just given; it is instead to suggest that mere family resemblance does not fully indicate how truly miscellaneous they are.

In a quite different context, and several years later, Wittgenstein did say: 'Mathematics is, then, a family; but that is not to say that we shall not mind what is incorporated into it' (VII, §33a, 399). But this sentence cannot be read out of context. The 'then' indicates it is part of a larger train of thought.

Earlier in the same paragraph we have a clause I much like, and abuse by repeating out of context: '– For mathematics is after all an anthropological phenomenon.' The 'For' indicates that the chain of thought does not begin with that observation either. The entire paragraph needs to be placed in *its* context within fragment VII. It begins in the middle of an internal dialogue, following on from §32:

> But in that case isn't it incorrect to say: the *essential* thing about mathematics is that it forms concepts? – For mathematics is after all an anthropological phenomenon. Thus we can recognize it as the essential thing about a great part of mathematics (of what is called 'mathematics') and yet say that it plays no part in other regions. This insight by itself will of course have some influence on people once they learn to see mathematics in this way. Mathematics is, then, a family; but that is not to say that we shall not mind what is incorporated into it (VII, §33e, 399). (I use letters to label paragraphs within a remark, so this is the fifth paragraph of *RFM*: §33.)

Wittgenstein was much attracted to the idea that mathematical proof affects the concepts used in the proof, so that in some sense it creates new concepts. It creates new criteria for the application of a concept. That topic was perhaps the one that most exercised early readers of the *RFM*, including myself (e.g. Hacking 1962).

This is not the occasion to explore the curiously complex §33, or to call up the internal dialogue within which it is embedded. We do note the tentative switch to 'what is called mathematics' from 'mathematics' – together with the fact that one participant in Wittgenstein's internal dialogue was looking for 'the *essential* thing about mathematics'. I doubt that there is a single short paragraph written by any notable philosopher that simultaneously displays what look like 'nominalist' and 'essentialist' temptations so explicitly.

14   Aside on method – on using Wittgenstein

We have just noticed a trivial difference between the way in which Felix Mühlhölzer and I understand Wittgenstein's use of the German words 'ein BUNTES *Gemisch*'. Mühlhölzer is the author of a valuable 600-page *Commentary* on Part III of the *Remarks on the Foundations of Mathematics* (78 pages). His book may variously be described as interpretation, history of philosophy, historical scholarship, or simply as commentary. In what follows I shall often use Wittgenstein's words, but I am *not* engaged in interpretation, historical scholarship, or commentary. I describe myself as a philosopher who reads Wittgenstein with some care, learns from what he reads, and incorporates what he has read, in his own idiosyncratic way, into his own philosophical thoughts. But I am not engaged in commentary.

Commentary is an enormously important part of philosophical activity. There are plenty of analytic philosophers who disparage the history of philosophy, but I am not among them. I learn a lot from other readers of classic texts and pay a good deal of attention to the commentators. But when I use Wittgenstein in thinking about mathematics, I am not interpreting him, except in the tedious highbrow sense in which any way of understanding anything anyone says is 'interpretation'. Happily I was able to read him – simply read him – before there were any commentaries or even reviews of his thoughts about maths.*

Hence I do not really care whether or not I get Wittgenstein right (though hubris makes me believe that I usually do!). I say only that it is of little significance to what I am doing whether I rightly understand the text. Of course I regret it if I misunderstand him, and hope to have my errors corrected. But if a misunderstanding has enabled a new line of philosophical reflection, that is an excellent side-effect of a text. To repeat, I am a philosopher who reads Wittgenstein with care and tries to use his words with respect. This may lead us astray from *Wittgenstein*, but not necessarily from *philosophy*.

That said, I shall from time to time offer unusual interpretations of the words of other canonical philosophers: for example, my proposal about Descartes, God, and no foundations in §1.31. These are not intended as serious contributions to scholarship but, as we might say, lateral thinking.

15 A semantic answer

Saul Kripke's (1980) theory of rigid designators changed the philosophical landscape of names forever. He taught that common names (such as 'gold' and 'horse') have a logic similar to that of proper names (such as 'Bismarck' and 'Obama'). One can imagine a noun denoting gold, or horses, coming into being by first pointing to an example. That would be a naming event rather like the baptism of Otto or Barack. Then there is a history of the use of the common name from then to now, just as there is a history of the usage of the proper names. This is just a picture, but it helps convey the idea that what gold *is*, is the kind of stuff originally used as an exemplar, and that its name is *directly* associated with it, and not, for example, by way of a description.

The name 'cholesterol' has a genuine history of this sort. Michel-Eugène Chevreul (1786–1889) isolated some interesting stuff, and in his first publication on the subject wrote: '*I give the name cholesterin to the crystallized substance in human gallstones* – from the Greek words for "bile" and "solid"'

(Chevreul 1816: 346, added emphasis). That is as close to a baptism as you will find in the annals of science. (Later, when it was realized that the stuff is, in chemical terms, an alcohol, the name was modified to 'cholesterol'.) Chevreul had a good guess at the chemical composition of the stuff, about which we have gone on to find more and more – work that has earned four Nobel Prizes shared between six people (H. O. Wieland, 1927; A. Windhaus, 1928; K. Bloch and F. Lynen, 1964; M. S. Brown and J. L. Goldstein, 1985). On the theory of rigid designation, 'cholesterol' denotes *this* stuff, to which Chevreul first drew attention, and whose chemical identity is now quite well understood, but whose properties are still being elucidated. My point, however, is somewhat ironical in intent. This sort of baptism is not the norm, but is in fact very, very rare. In the vast post-Kripkean literature, no one else, to my knowledge, has produced a similar documented example of a 'baptism' of a now-familiar substance.

Could mathematics be like cholesterol? The usage of terms denoting mathematics goes back a very long time in history, quite probably long before the time of Thales, in Babylonia and Egypt, and, only a bit later, in China. Had there been writers of dictionaries over the course of all history, we would have had a far greater miscellany of definitions than that with which I began this chapter. Nevertheless, on the picture of rigid designation, there would be something at which all the users of names for mathematics were unwittingly pointing – *mathematics*.

To avoid misunderstanding, I must make plain that I know of no one who has asserted that the name 'mathematics' is a rigid designator, and that I am not attributing the idea to anyone, and certainly not to Saul Kripke, who invented the theory of rigid designation. I am saying only that the picture could be put to use in this way.

Unfortunately doing so does not tell us what makes mathematics mathematics. Chevreul began to tell us what makes cholesterol cholesterol, and all those Nobel Prizes honoured men who told us more about that. Do we need philosophical analysis to do work comparable to chemical analysis in order to answer our question? Or should we turn to mathematicians?

16   More miscellany

William Thurston (1946–2012, Fields Medal 1982) wrote a piece that should, in my opinion, be compulsory reading for all philosophers who think about maths: 'On proof and progress in mathematics' (Thurston 1994). I say this not because he is right in most of what he says (though I do think he is), but because he dispels many misconceptions of mathematics

which were current fifty years ago, many of which are still in circulation. He
has an important, and novel, answer to the question, 'what makes maths
maths?' But first, two of many other things one can learn from his piece.
One has to do with the miscellany of mathematics. The other bears on
changing conceptions of mathematical proof. After sketching those
thoughts, we return to our question of what makes maths maths.

The first thing that a mathematics student learns on entering college
(and often before) is the differential calculus. In order to differentiate, you
first learn about derivatives. But these are not taught in a unique way,
because there are many ways to conceptualize a derivative. There are
many different stories, pictures, analogies, and tricks of the trade that
build up understanding.

Thurston (1994: 163) lists seven different ways in which to think of
derivatives (and there are more). He labels his seven (1) Infinitesimal.
(2) Symbolic. (3) Logical. (4) Geometric. (5) Rate of instantaneous speed.
(6) Approximation. (7) Microscopic. Anyone familiar with the calculus will
recognize what he means. Thus (4) means thinking of the derivative as the
slope of a line tangent to the graph of a function, so long as the graph does
have a tangent. (5) means the speed of $f(t)$, where t is time, at any given
moment of time.

Many readers will remember thinking in one or more (or all) of these
ways as knowledge and practice grew. I myself never thought in terms of (7):
'The derivative of a function is the limit of what you get by looking at it
under a microscope of higher and higher power.' I never had that picture,
but it could be a helpful metaphor.

At one stage in his thinking about mathematics, one of Wittgenstein's
internal voices would probably have proposed (in what I call his 'try-out
mode') that in adding (7) to my roster of ways of understanding derivatives,
I have changed my concept of a derivative. And if we do not attach too
much logical, 'Fregean' weight to that, it is a perfectly plausible thing to say.
But it is better, in the contemporary English of analytic philosophers, to say
I have augmented my conception of the derivative, but I have not obtained
a new concept.

17   Proof

Thurston has many wise words to say about proof, a subject to which we
shall often return. In particular he recounts how conceptions of proof have
changed, not only from the days of Leibniz and Descartes, but also during
his own working lifetime. I should mention that his essay was the longest

response to a polemical essay by Arthur Jaffe and Frank Quinn (1993), about mathematics and theoretical physics. A number of distinguished mathematicians responded. Brown (2008: 207–17) summarizes the debate under the heading 'the math wars'. I avoid the image of war, for war is *monstrous*, and the debate was civil, but there was indeed a debate, well explained by Brown, to which Thurston's paper was a contribution.

One thing of which Thurston reminds us is that proof, like any other kind of evidence, comes in degrees. Fifty years ago it was taken for granted by most mathematicians, logicians, and philosophers that demonstrative proof is a yes-or-no matter. Either a proof is valid, or it is fallacious, and that's that.

Wittgenstein, doing his thinking in the 1930s and 1940s, was obsessed by proof (which, as I have said, is the most frequently used substantive word in the *RFM*). He had a far wider range of examples of proofs (or maybe we had better say 'proofs') than anyone else did in those days. Nevertheless proof, in his writing, seems to be a yes-or-no matter.

The same is pretty much true, in a different way, for Imre Lakatos' (1976) *Proofs and Refutations*, most of which was initially published in 1963–4 (cf. §1.27 above). It really shook up the prevalent conception of proof. Lakatos illustrated how many proofs of famous results went through endless revisions, because counter-examples were discovered. But that did not mean that they were proofs only to a degree. It meant that they were fallacious, and had to be corrected by various strategies to which he gave aphoristic names.

The conception of proof as conferring certainty, widespread in the period when Wittgenstein and Lakatos wrote, i.e. the mid-twentieth century, has changed, in part due to developments in mathematics itself. Proofs have, to put it crudely, become longer and longer, so that it is not possible for a single human to grasp them in their entirety. You might think this just shifts the burden from cartesian proofs to leibnizian proofs, checked by computer. Today that brings with it the question of the error rate of a given computer, a topic that has not been much discussed by philosophers, but which has become a branch of mathematics in its own right.

18 Experimental mathematics

There *has* been a lot of discussion, by both mathematicians and philosophers, about proofs checkable only by computer, and about computer-generated proofs. The prime example to which philosophers usually turn is the four-colour theorem. The proof goes back to Appel and Haken (1977) continuing with Appel, Haken, and Koch (1977). Yes, much of the proof was exhaustion by cases of a list of alternatives uncheckable by human hands. But what has not been emphasized is that much of the immediate discussion was less about the work done by computers than about a very long hand-written series of calculations that verified that the list of alternatives was exhaustive. In fact the proof was neither cartesian nor wholly leibnizian.

Far more important, in my opinion, is the advent of experimental mathematics, with its own journal, *Experimental Mathematics*, founded in 1992. To avoid misunderstanding, it was emphasized from the start that, first, mathematicians have been doing experimental mathematics forever, doodling with pencil and paper, or playing around with figures drawn in the sand. Second, this was to be a journal of 'pure' mathematics. One point was to make plain that the journal was not dedicated, for example, to the booming activity – nay, industry – of simulating experiments in the material world. That depends, of course, on the constant use of mathematical reasoning both in modelling the micro and macro universes around us and in designing programmes in which the models are embedded. But it was not the mission of the new journal to contribute to that kind of work.

Thus far, most philosophers seem to have discounted experimental mathematics as nothing new (van Bendegem 1998, Baker 2008). They perhaps do not attend to the way in which computers are such a powerful tool for mathematical exploration. My consciousness was first raised in the mid 1980s when a somewhat eccentric Paris topologist* would stay for a week or so in our home in Toronto, lugging an enormous Mac, which he would set up in the basement and to which he was glued all night, not for proof-construction, but for experimenting on topological conjectures.

How about this for a sentence destined to rouse hackles? – Experimental mathematics provides the best argument for 'Platonism' in mathematics: that is, the idea that mathematics is just 'out there', a given. We explore it with many tools, including pencil and paper, and now computers. But contrary to many philosophers, this does not leave everything the same, not to worry. The journal *Experimental Mathematics* was founded by David Epstein.* As he wrote in an email (2 September 2010), these are 'things that no-one dreamt of when we were students', namely in the late 1950s. And of course to say that experimental mathematics provides the best argument that mathematics is 'out there' is not to say that it suggests that maths is out there 'in some non-physical realm'. I use 'out there' in scare quotes to reflect an experience of 'givenness' that some people have, and which a few mathematicians actually express using those words. I do not use it to practise transcendental geography.

Sometimes what is found out by experimental mathematics is quickly replaced by a deductive proof. Some of those deductive proofs are old-fashioned proofs that make sense, while others are themselves long and not very memorable exercises, perhaps only checkable by computers. Here there is a real division of attitude. Some mathematicians regard computers merely as tools of discovery or as search machines for counter-examples. After discovery comes justification. Reichenbach's notorious contexts of discovery and justification, first articulated in 1923, which Kuhn had, for many readers, demolished in 1962, seems once again to be pertinent, at least in mathematics!

Others think that the old proof-oriented attitude is obsolete. I do not take sides, but I do think of proof itself as a concept that has been evolving since 'Thales or some other' discovered how to make proofs. I cannot guess how proof will evolve tomorrow, and I take no stance on how it 'ought' to evolve. There may be a touch of déjà vu all over again. The 'leibnizians' and 'cartesians' are jousting on new terrain, fast computing, using old lances. Except Leibniz might be saying, 'I told you so!'

Mathematicians themselves differ, and their attitudes evolve in different ways. Timothy Gowers (Fields Medal 1998) is the author of the wonderful *Mathematics: A Very Short Introduction* (2002), in which he writes, 'My own view, which is a minority one . . . is that over the next hundred years or so [computers will be] eventually supplanting us entirely' (2002: 134). But, as he continues, 'Most mathematicians are far more pessimistic (or should that be optimistic?) about how good computers will ever be at mathematics.' Vladimir Voevodsky (§1.22) is not among the pessimists about the future role of fast computers in mathematics.

Incidentally, and perhaps this is relevant, I have encountered very few working mathematicians who take Wittgenstein seriously. Gowers writes, 'Anybody who has read this book and the *Philosophical Investigations* will see how much the later Wittgenstein has influenced my philosophical outlook and in particular my views on the abstract method' (2002: 139f). We shall return to Gowers in §§6.14–26, where he serves as my archetypal anti-Platonist.

Gowers also preaches the value of large-scale collaborative maths using the internet. He has perhaps the most extensive and sustained personal mathematical blog in existence. He presents his own conjectures and explorations, and invites the whole wide world to contribute ideas. In 2009 he founded the Polymath Project for collaborations. Successful results in combinatorics have been published under the collective name, D. H. J. Polymath. There has been the mathematics tearoom for a century; is this to be the new global tearoom?

19   Thurston's answer to the question 'what makes?'

Thurston also has a few important words about what makes mathematics mathematics:

> Could the difficulty in giving a good direct definition of mathematics be an essential one, indicating that mathematics has an essential recursive quality? Along these lines we might say that mathematics is the smallest subject satisfying the following:
> - Mathematics includes the natural numbers and plane and solid geometry.
> - Mathematics is that which mathematicians study.
> - Mathematicians are those humans who advance human understanding of mathematics.
>
> In other words, as mathematics advances, we incorporate it into our thinking. As our thinking becomes more sophisticated, we generate new mathematical concepts and new mathematical structures: the subject matter of mathematics changes to reflect how we think. (Thurston 1994: 162, bullets in original)

Such a consideration 'brings to the fore something that is fundamental and pervasive: that what we [mathematicians] are doing is finding ways for *people* to understand and think about mathematics' (his italics).

We might add a converse to his final sentence: the subject matter of mathematics changes to reflect how we think, and how we think about mathematics changes as the subject matter changes. Notice that we could say *this* about any science: recall what molecular biology has done to biology. But I doubt we would define biology recursively in the way that Thurston defines mathematics.

Thurston meant his three bulleted points as almost literally a recursion as understood by logicians. There is an agreed starting assertion, the basis of the recursion. In this case, number theory and geometry are mathematics. This rightly takes for granted what I found puzzling in Chapter 1 – that arithmetic and geometry should be part of the same discipline – but does not commit to what I found astonishing – that they are so intimately and profoundly intertwined. Then we introduce people, namely mathematicians. What they do is what determines what mathematics is. Likewise for number theory, which I often refer to as arithmetic. There is a vast amount of nutty numerology, such as counting the number of syllables in a certain chapter in a favoured Holy Book. But that is not what mathematicians do, at least in public. (Many, from Newton down, do a lot of strange figuring in private, so the demarcation is not so sharp as a rationalist would like.)

Many readers will find this deeply unsettling – surely mathematicians do what they do because it is mathematics? Or rather, there are certain features of what they do which attracts them, and those features are what make mathematics mathematics. Yet that is the type of statement Thurston is implicitly rejecting. There is no set of features that determines what counts as mathematics. Certainly no set of necessary and sufficient conditions. Many philosophers will not be troubled by that, and say simply that what counts as mathematics forms a family, but others may feel that is a cop-out. We are actually not very good at characterizing family resemblances. With human families, which Wittgenstein used as a model, we can point to the nose, the characteristic gait, or the wry smile, etc. I will go on to point at features shared by this or that group of examples, but it is not a satisfying exercise.

In §2.13 we quoted Wittgenstein: 'Mathematics is, then, a family; but that is not to say we shall not mind what is incorporated in it.' And who is this regal 'we'? Is what makes the family what mathematicians do? (And do they do maths because of the personality quirks mentioned in §2.9 above?)

20 On advance

'Mathematicians', Thurston said, 'are those humans who advance human understanding of mathematics'. He did not say, advance human *knowledge* of mathematics, but *understanding*. This includes new theorems, more proofs, for sure, but also new concepts, new kinds of proof, new analogies, and new connections between fields of research that began with very different motivations. To repeat the sermon: mathematical advance is not just a matter of proving new theorems, but also of new ideas, new techniques, new questions, new ways to prove, and more generally new ways to investigate – and, since advance is ragged, abandoning some ways of thinking as infertile.

Thurston cites a visible example from his own research – by visible, I mean that a silent movie of Thurston at work would have shown the use of a new tool. By 1994, when he published his essay, he was spending more and more time using his computer to investigate mathematical structures – not to make 'computer-generated proofs' which have attracted philosophical attention, but for mathematical exploration, as just described. Computers, to repeat, have radically changed the day-to-day lives of some mathematicians, while other mathematicians disdain them.

I use Fermat's fact, about primes of the form $(4n + 1)$, as an arbitrary but convenient example of an easily understood proposition that is not in the least obvious, and which has a relatively elementary proof. Philosophers of mathematics, including the exceptional Wittgenstein and Lakatos, have focused their attention on such theorems and their proofs. Thurston reminds us that theorem-proving is only one part of mathematical activity.

It is also an important part. The growth of knowledge of mathematical facts is central. Yet medals and prizes do not reward new theorems as such, but rather new proof ideas which pass beyond the particular facts they are used to establish. If we return to the older picture, and think of mathematical knowledge as something that advances by the accumulation of theorems, there is still a question to ask. Is the advance in a pretty definite direction comparable to the development of an insect from pupa to butterfly? Or is the future of mathematics very open, and dependent on what mathematicians happen to do, much as a language is the work of its users? Will it confirm Gowers' expectations? Or will they shrivel and die?

21 Hilbert and the Millennium

It is a mark of changing times that in 1900, one man and his school, David Hilbert, set the problems of the century to come. One hundred years later, a committee was called for. Mathematics had become a globally collaborative (as well as competitive) enterprise. The committee was formed under the aegis of the newly founded Clay Mathematical Institute, established by a Boston businessman and his wife, Mr and Mrs L. T. Clay. It supports many projects and people, but was publicly noticed for establishing the Millennium Prizes, each worth a million dollars for solving an outstanding problem. The Clay Institute even re-enacted Hilbert's 1900 address in Paris, with Alain Connes announcing the seven problems in Paris in 2000, and a popular lecture by Timothy Gowers. I mention this only because I use Connes and Gowers in Chapter 6 as exemplary of Platonism and anti-Platonism among mathematicians. Videos of Gowers, and of Michael Atiyah and John Tate describing the Millennium problems, are found at www.claymath.org/annual_meeting/2000_Millennium_Event/Video.

The problems divided, very roughly, into two algebraic problems, two topological problems, two problems in mathematical physics, and one problem in the theory of computation. There are numerous popular expositions of these problems; for an intelligent but easy-to-understand sketch, I recommend Cipra (2002; ignore his flippant title). Because of the ready availability of such summaries, I will not explain the problems, and instead merely name them. I use them only to illustrate what is older and what is newer. Otherwise it is totally distorting to focus on topics chosen by price tag.

The Millennium list was in many ways patterned after Hilbert's list. Some of his twenty-three have been solved, some are still open, and some were not stated precisely enough for an agreed definite conclusion to be reached. Grattan-Guinness (2000) counts three that, at the time, were probably not clear enough to be definite solvable problems. Two more are best described as groups of problems, and a further five divide into two fairly distinct problems.

The first two in Hilbert's list are familiar to philosophers. No. 1 was the continuum hypothesis and a well-ordering principle, shown by Gödel to be consistent with standard axioms of set theory, and by Paul Cohen in 1963 to be independent of them. No. 2 was the task of proving the consistency of 'arithmetical axioms'. Grattan-Guinness suggests that Hilbert's statement was perhaps not precise enough, in 1900, to be a clear problem. Everyone now knows, thanks to Gödel in 1931, that a consistency proof is impossible within any consistent axiom system adequate for recursive arithmetic. In parenthesis, thereby hangs a philosophical question that only a few philosophers have addressed. During the 1930s, Gerhard Gentzen (1909–45), a student of Paul Bernays and later Hilbert's assistant in Göttingen, gave (to say the least) interesting evidence that first-order arithmetic is consistent. But it requires transfinite induction up to a 'small' infinite ordinal (namely ε_0). Is Gentzen's argument adequate to the task? Does it leapfrog Gödel?

We have good evidence that late in his career Gödel himself was still pondering Gentzen's work. We commonly say that Gödel's second incompleteness theorem brought an end to Hilbert's programme. C. G. Hempel used to tell an anecdote about his student days. In Berlin he attended John von Neumann's seminar on Hilbert's programme. One day von Neumann walked in waving some pieces of paper. 'I have received a letter from a young man in Vienna. This seminar is cancelled.' But that is not how Gödel himself came to see things. Toledo (2011: 203) reports a conversation of 13 June 1974 in which Gödel said that: 'Hilbert's program was completely refuted, but not by Gödel's results alone. That Hilbert's goal was impossible became clear after Gentzen's method of extending finitary mathematics to its utmost limits.' On 26 July (2011: 204) he continued with a provocative gloss on this observation, and the suggestion that we still have not digested Gentzen's approach.

Some of the seven Millennium problems were on Hilbert's list, and at least one could not possibly have been. The number-theoretic Riemann hypothesis was one of two problems about prime numbers that Hilbert had as fourth in his lecture and no. 8 in the list of twenty-three. Bernhard Riemann (1826–66) proposed his hypothesis in 1859, but it turns out to have astonishing ramifications, and remains unproven. The Langlands programme (§1.15) is strongly connected to it. The Poincaré conjecture, which is essentially topological, is also old, as the name implies. Henri Poincaré stated it in 1904. It was proven by Grigori Perelman in 2002. He became rather famous, first for refusing the Fields Medal in 2006, and then, in March 2010, for refusing the million dollars that the Clay Institute judged he had earned.

The Hodge conjecture in algebraic topology was proposed by William Hodge (1903–75) in 1941, although the ideas are older than that. It did not become noticed until 1950, and indeed in its first form was actually refuted and then revised in 1969.

The first physics problem, in this case named after Claude-Louis Navier (1785–1836) and G. G. Stokes (1819–1903), is from good old nineteenth-century fluid dynamics, which had in its essentials to be learned by every twentieth-century undergraduate specialist in mathematics and physics. But in a deep sense no one quite understands why water does flow, or the circumstances in which it does not flow. Do the Navier–Stokes equations always have a smooth solution? (These equations recur in §5.28 below.)

In contrast, the second physics problem, Yang-Mills, makes sense only in the context of well-developed quantum mechanics, namely quantum chromodynamics and more generally gauge theory. It derives specifically from the work of C.-N. Yang (b. 1922, Nobel Prize 1957) and R. L. Mills (1927–99). To ride as roughshod as possible, the task is to explain why matter is as it is, and sometimes is not (there are 'mass gaps'). This problem was not really intelligible until the 1950s, and is still not properly understood.

Now notice two of the seven where the emergence of modern computing has made a serious difference, but in two very different ways. In one case, a conjecture is made plausible by using computers to test countless special cases to try to find a counter-example – and failing. That's the familiar use of computers for exploration. In the other case, the problem is deemed important precisely because, although it is stated as an abstract question in 'pure' mathematics, it is very close to problems in real-time computing.

The Birch and Swinnerton-Dyer conjecture of the 1960s, like the earlier Hodge conjecture, derives from work earlier in the twentieth century. But it became 'serious' when Birch (b. 1931) and Swinnerton-Dyer (b. 1927) (and their students!) checked it for many very large primes, using the rather early but powerful EDSAC computer at Cambridge University. If we think that exploration by machine is what made the conjecture plausible, it could be classified as necessarily post-World War II. But the advent of computation, and then fast computation, is *not* what made the question interesting. It was just experimental mathematics with a new tool.

In contrast, the P versus NP problem, although fundamental in its own right, gains its cogency from real-time computing. It was a natural for the new field of complexity theory when it was advanced by Stephen Cook (b. 1936) in 1971. Could the P=NP problem have been formulated earlier by, say, Turing?

Whatever 'could' have been thought, it almost certainly was not. Stephen Cook* says that 'in 1956 Gödel wrote a letter to von Neumann in which he hinted at the notion of NP-completeness. Also there were Russian mathematicians talking about "perebor" (search problems) in the 1950s. But I am not aware of anyone formulating concepts akin to P vs NP before then' (email 22 August 2010). Note, by the way, that P=NP is the only one of the seven Millennium problems that is *not* named after a person, even though Cook probably *did* invent the concepts and frame the problem. This may suggest a corollary to Stigler's law of eponymy. Stephen Stigler is a mathematical statistician and historian of probability ideas. His law, in short form, reads: 'No scientific discovery is named after its original discoverer' (Stigler 1980). It is usually named after a later 'discoverer'. (Stigler notes that Stigler's law applies to itself, for it was first implied by Robert Merton.) The corollary is that if everyone agrees on an original discoverer, the law will not be named after anyone.

22 Symmetry

The previous section illustrates both inertia and novelty. Even when the same problem is posed, it is in a totally different setting from its early appearance. That's 'advance'. Now I shall draw attention to a curious phenomenon. For the past few decades, symmetry has been a buzz-word. One might even argue that the very possibility of mathematics derives from our innate sense of symmetry. We can now read considerations of symmetry back into the earliest deep mathematics, such as the Platonic solids, each of which is notable and indeed intrinsically interesting because of its symmetries. Hon and Goldstein (2008) tell convincingly, in the course of an immense book, the story of how symmetry began to be an important mathematical idea only at the time of Legendre. It became the central concept of group theory. But it seems not to have played a major part in Western mathematical or scientific awareness until, let us say, the time of the French Revolution.

This already creates a dissonance, for surely people were aware of symmetries forever? But there is a further dissonance. We now see group theory and symmetries as mates: any group defines a symmetry, and any symmetry defines a group. We would expect these to be at the forefront not only of research but also of general mathematical education, yet here is a passage by Joe Rosen (1998: vii) writing about the first edition of his book *Symmetry Discovered* in 1975. He said he wrote it because he had discovered a desert: 'between the coastal plain of a couple of children's books and the heights of Weyl's *Symmetry* [1952] there was nothing but barren wilderness'. Today, Amazon USA lists some five thousand titles about symmetry, many of which really do bear on mathematics or physics, and many of which are semi-popular expositions, e.g. Ian Stewart's *Why Beauty is Truth: A History of Symmetry* (2007). There is the fascinating, readable, but immensely deep *The Symmetries of Things* (Conway, Burgiel, and Goodman-Strauss 2008). That is perhaps the most profound coffee-table book ever published.

A thousand flowers have bloomed, it appears, in the desert that Rosen encountered before 1975. One reason is that symmetries have been immensely important to mathematical physics, leading to the resurgence of pythagoreanism (Steiner 1998, Hacking 2012b). Here we have advance of a different sort, in part through a change in setting, as discussed in §4.5. More on symmetry in §5.20.

23   The Butterfly Model

Thurston's observations give a different cast to our question, 'What makes maths maths?' Mathematics, as practised today, is itself the temporary end-product of a historical chain of events. As long as there are ingenious mathematicians at work, it will probably continue to develop, but in ways that can at most be dimly foreseen. Mathematics, from this point of view, is like a living organism which grows, and which will be perceived differently over the course of time. Hence my second epigraph, due to Lakatos (1976), who said exactly that: 'Mathematics, this product of human activity, "alienates itself" from the human activity which has been producing it. It becomes a living, growing, organism.'

That is not the end of the metaphor. There are two organic models of mathematical advance to which I shall attach cute names.

Real organisms – biological ones – change as they grow from egg to infant to teenager to adult to geriatric, as recorded in Shakespeare's seven ages of man. More dramatically (if possible) is the metamorphosis from egg to larva to pupa to butterfly. These changes are inevitable given enough nourishment, barring accident, disease, or failure to thrive. 'Failure to thrive' is an approved diagnosis in paediatrics, but could nicely apply to a degenerating science that was getting nowhere.

An organism is teleological: that is, it develops from almost nothing towards a final end. In the case of animals, insects, etc., it grows towards a mature creature. The full cycle for organisms ends in death, which is more inevitable than any change in the course of a life, but let us use a maturing organism as one simile for mathematics, with the option that there is no final end. A central feature of this teleological picture is that the course of maturation is pre-determined, barring intervention or failure to thrive. Is mathematics like that?

We could call this the maturation model of mathematics, or the teleological model, but for drama, and to allow for possibilities of metamorphosis, let us call it the *Butterfly Model*.

24 Could 'mathematics' be a 'fluke of history'?

Thurston implies that what we call mathematics is a growing thing, so long as mathematicians carry on creatively and do not run into complete dead ends everywhere. Their work – and it should be regarded as *work* in something like the nineteenth-century labour theory of value – leads us not only to new results, but also enlarges the ways in which people think about the subject. We cannot define what we call mathematics because it is more like an organism than a static object. But we could still have the Butterfly Model of inevitable development.

There is a more radical view of this historical process. Every year, the mathematician Doron Zeilberger publishes a few 'Opinions' online in order to shake up his colleagues. They do disturb and annoy some people, but I agree with many of his dicta. Maybe they are less shocking to philosophers than to mathematicians.

> *Our* mathematics [he posted on 25 April 2010] is an accidental outcome of the *random walk* of history, and would have been very different with a different historical narrative. Even if, for the sake of argument, there is an 'objective' mathematics out there, independent of us (or of the creatures in the fifth planet of star number 130103 in Galaxy number 4132, who are far smarter than us), whatever tiny fraction of it that *we* (or even our smarter colleagues from that galaxy) could have discovered, is entirely a fluke of history. (Zeilberger website, his italics)

To propose that our maths is the result of a random walk, a fluke of history, is not to suggest that (recalling Wittgenstein's words) we shall not mind what is incorporated in it. It is only to assert that what we are willing to count as mathematics is a consequence of historical accidents and was by no means inevitable. We need a model different from the pre-determined maturation of the butterfly.

25 The Latin Model

Is the future of mathematics very open, not in any way pre-determined and largely dependent on what mathematicians happen to do? Is the organism that is mathematics more like a language, which evolves in no particular direction? We don't want the metaphor to suggest that 'anything goes': that mathematics could go any which way. Although the future of a language is very open, there are also a lot of constraints. According to theories derived from Noam Chomsky, there is an underlying universal grammar which is part of the human inheritance. Philologists and empirical linguists more modestly find certain regularities in the evolution of both phonetics and grammar, regularities sometimes dignified by calling them laws.

All I want as an alternative to the Butterfly Model is the idea of a language developing in no particular direction, but subject to constraints. It is the picture of Latin evolving into Spanish, a spontaneous and unplanned historical development. Under other circumstances, it could and did evolve into Italian or Romanian or Esperanto, and it could have evolved into any other Romance language, including many possible ones that never in fact arrived.

To firm up the present metaphor let us not think of the actual history of Romance languages, but of fiction: something a bit more like the actual history of Chinese. Imagine that the only descendant of Latin, spoken all the way from Bucharest to Valparaiso, was Spanish. Various dialects, to be sure – Castilian, Mexican, Argentine – but what is still taken to be just one Spanish. In this fiction it was by no means inevitable that Spanish would have evolved from Latin, even if it was the only way that Latin did evolve. Perhaps Spanish speakers would be unable to envisage alternatives, but there could have been other Romance languages, such as Italian, as we outsiders to the fiction know full well. It just so happens, in the fiction, that there are none. Let us call this the *Latin Model*. That may be something like Zeilberger's image of mathematics.

Note that, 'for the sake of argument', Zeilberger does not deny there might be a definite '"objective" mathematics out there, independent of human thought' – though obviously he does not think there is such a thing.

(Note the 'out there', a locution to which we shall return. Out where?) Note that in the final sentence of the quotation he uses the language of discovery. He says only that our discoveries, and indeed the carving out of the continents in which the discoveries could take place, is the consequence of a history that is not pre-determined. The metaphor of exploration is often used. Europeans exploring North America came from the East, and inevitably got to the St Lawrence before the Mackenzie River system. Much earlier, Asians coming from the West discovered the Fraser or Columbia systems before the Mackenzie. But in the end, most people suppose, there was just one continent to discover.

Zeilberger is not, in the passage quoted, imagining that a different walk through mathematical history could have yielded results formally inconsistent with results at present established. Subject to the usual proviso: *unless*, when the results were brought together, it was possible to see that one group of results was fallacious, an over-generalization, or what-not.

26 Inevitable or contingent?

To speak of flukes and random walks is the exaggeration of a controversialist who has just woken up to a fresh way of looking at his field. What a good wake-up call! But some distinctions are called for. Zeilberger is saying that the body of knowledge we recognize as mathematics is contingent on a long series of historical accidents. Mathematical *results*, in the form of theorems or constructions, are, on the usual understanding, the very opposite of contingent; they are not contingent on anything. Philosophers call them necessary truths. Zeilberger is not taking issue, *here*, with that traditional notion.

When Zeilberger, for the sake of argument, does not deny that there is an objective mathematics out there, independent of us or any other intelligent creature, he reminds us that his view of actual mathematics as a random walk is completely unrelated to the traditional or the current '-isms' in philosophies of mathematics. None of the current favourites – platonism, nominalism, and structuralism, say – has or implies any view about whether current mathematics is the contingent result of a random walk, the Latin Model, or, inevitably, the Butterfly Model. The division is rather between all those philosophies on one side, and a 'social construction' attitude on the other side. We return to these matters in §§4.2–5, where the two types of models will be put to use. As this book continues, it will become clear that about important things I tend to favour the Latin model.

27 Play

Our discussion has been inconclusive. That is perhaps how it should be. Rather than refining Thurston's recursion with serious analysis it may be more thought-provoking to reflect on the remarks of a mathematical entertainer, a puzzle-maker – none other than the Dudeny who may have suggested the problem of squaring the square, discussed in §1.28.

> The history of the subject [of mathematical puzzles] entails nothing short of the actual story of the beginnings and development of exact thinking in man. The historian must start from the time when man first succeeded in counting his ten fingers and in dividing an apple into two approximately equal parts. Every puzzle that is worthy of consideration can be referred to mathematics and logic. Every man, woman, and child who tries to 'reason out' the answer to the simplest puzzle is working, though not of necessity consciously, on mathematical lines. Even those puzzles that we have no way of attacking except by haphazard attempts can be brought under a method of what has been called 'glorified trial' – a system of shortening our labours by avoiding or eliminating what our reason tells us is useless. It is, in fact, not easy to say sometimes where the 'empirical' begins and where it ends. (Dudeny 1917: 1)

We could write a whole chapter on the final sentence (indeed you can read Chapter 5 as just such a chapter!). Let us, however, direct our thoughts to the first sentence. Thurston began with the theory of numbers and geometry. Dudeny takes arithmetic back to counting, and solid geometry back to dividing an apple. But not everything that descends from them is mathematics! So let us call in G. H. Hardy, thinking about chess problems, which are just one type of puzzle.

> A chess problem is genuine mathematics, but it is in some way 'trivial' mathematics. However ingenious and intricate, however original and surprising the moves, there is something essential lacking. Chess problems are *unimportant*. The best mathematics is *serious* as well as beautiful – 'important' if you like, but the word is very ambiguous, and 'serious' expresses what I mean much better. (Hardy 1940: 88f)

Steiner (1998: 64) maintains that chess and chess problems do not count as mathematics at all. By now we should be clear that the family that is mathematics does not have sharp boundaries, and that there may be reasons for including chess problems, and other reasons for excluding them. Hardy is not examining what mathematics is, but speaks of 'the best mathematics', and it is *serious*. He would not complain about the demand of the *Transactions of the American Mathematical Society*: 'To be published in the *Transactions*, a paper must be correct, new, and significant.' It goes on to say that 'it must be well written and of interest to a substantial number of mathematicians'.

28 Mathematical games, ludic proof

Dudeny called the book from which I have quoted *Amusements in Mathematics*. Older readers will recall Martin Gardner's (1914–2010) column 'Mathematical Games' in *The Scientific American* (1956–81). It was in many ways very serious indeed, and Dudeny was serious in many of the same ways. There are paths that lead on from mere amusements to things that Hardy might count as serious. The example of squaring the square is a case in point. Tutte's own (1958) account of the events was published in Gardner's column. The problem began in part with Dudeny's puzzle of Lady Isabel's Casket (§1.28). It turned into moderately deep questions with applications to electrical analysis. And Hardy's collaborator Littlewood used the proof, of the impossibility of a cube dissection, as an exemplar of good proof. A silly puzzle evolves into fairly good maths.

John Conway (b. 1937) is only one outstanding example of a creative mathematician fascinated by mathematical games. (A trifle more on Conway in §3.10 below.) His game *Life*, inspired perhaps by John von Neumann's idea of a self-replicating computer, was first published in Gardner's column in 1970. It has an astonishing number of properties, including being a Turing Universal Computer, and seeming to simulate all sorts of organic processes. Daniel Dennett has repeatedly used it to illustrate philosophical theses in *Freedom Evolves* (2003).

But all that is just icing on the cake, still leaving us with the idea that the games are mere froth coincidentally associated with the solemn activity that is mathematics. Mathematics is so often presented, almost especially by its philosophers, as something austere, august, stately, and humourless, that we tend to forget how playful so much mathematics is. In the Dudeny version of the Thurston recursion, the basis for the recursion on which a definition of mathematics is to be found is solving puzzles. Not just to survive, as that ultimate plodder of a discipline called evolutionary psychology would teach, but because it is fun. Granted, it becomes what we call mathematics when it becomes what Hardy called serious, and what others call rich, while others would point to applications in understanding and controlling the world.

To keep on with a string of vague words best not explained, puzzles lead to structures, and when they are rich they are mathematical. Rich: Leibniz's principle of sufficient reason gives a clue. We want the maximum of complexity from the simplest input. Leibniz, having a mathematical mind, thought that was the principle that God used when designing the universe. In a healthy theology, God was just having fun, and we are his playmates.

None of these key words are to be taken as definitive; they are brush-strokes for recalling some of the more general aspects of the motley that is mathematics. Here is one in particular. Reviel Netz is the foremost author-ity on Archimedes today. He is editing all of Archimedes, including the recently recovered palimpsest, the oldest surviving text of Archimedes, and the oldest surviving text with diagrams (Netz and Noel 2007). He calls his book focused on Archimedes, and on the style of the ancient mathemati-cians, *Ludic Proof* (Netz 2009). It emphasizes the role of play in ancient proofs. The title recalls Johan Huizinga (1872–1945) and his *Homo Ludens* (1949), with its famous thesis worked out in the 1930s: play is an essential element in the formation of human civilizations. Play is an essential element when we begin to ask how mathematics became possible. Play for its own sake, play as fun. (I regret that the very word 'fun' has been degraded by the media and shills for package holidays.)

We tend to think it is just a metaphor, even an insulting metaphor, when people say pure mathematics is just a game. Harsh Frege, inveighing against the formalists of his time, says in §91 of *Grundgesetze der Arithmetik* II: 'it is the applicability alone which elevates arithmetic from a game to the rank of science. So applicability necessarily belongs to it' (Frege 1952: 187). Stern Wittgenstein reiterated the judgement: 'It is the use outside mathematics, and so the *meaning* of the signs, that makes the sign-game into mathematics' (*RFM*: v, §2, 257). Could we imagine someone reversing this dictum? 'Mere applicability debases arithmetic from its status as a game and generator of puzzles, and reduces it to usefulness.'

One of the innumerable things that makes (some) mathematics mathe-matics is playfulness.

The thoughts pursued in this chapter will seem some distance from what is usually understood by the philosophy of mathematics. So now we turn to my title question, which is seldom if ever asked: why is there philosophy of mathematics *at all*?

CHAPTER 3

Why is *there philosophy of mathematics?*

1   A perennial topic

Why is the philosophy of mathematics 'perennial'? Perennial in the gardener's sense: not always flourishing, but reappearing, to flourish in almost every fit season. That is what I mean by asking: why is there philosophy of mathematics *at all*?

This is not the question of why the philosophy of mathematics is a thriving academic sub-speciality right now. There are a great many things about mathematics for philosophical minds to contemplate. Classic examples are the antinomies, introduced by Russell's paradox, Gödel's incompleteness theorems, the continuum hypothesis, and, more generally, intuitionism and constructivism. They alone provide ample sustenance for a speciality called 'philosophy of mathematics', and there is much more. Problems generated by computation, P/NP, for example (Stephen Cook's problem, §2.21). Is that a mathematical question or a philosophical question? But that's all twentieth-century, even if some of if it has Kantian roots.

If philosophical thinking about mathematics were confined to such matters, it would be no more and no less important than the philosophy of any other special science, driven by fairly recent puzzles about radically new results and events. It would be filed away in discreet academic locales, just like the philosophy of physics, the philosophy of biology, the philosophy of economics, or, less familiar but highly innovative, the philosophy of geography.

In fact, however, mathematics is the only specialist branch of human knowledge that has consistently obsessed many dead great men in the Western philosophical canon. Not all, for sure, but Plato, Descartes, Leibniz, Kant, Husserl, and Wittgenstein form a daunting array. And that omits the angry sceptics about the significance or soundness of mathematical knowledge, such as Berkeley and Mill, and the logicians, such as Aristotle and Russell. For brief case-by-case discussions of these figures and more, see 'What mathematics has done to some and only some philosophers' (Hacking 2000b). The writings of these men were not originally billed as 'philosophy of mathematics' – just philosophy – but retrospectively we recognize them as having contributed to, and even established, what we now call the philosophy of mathematics.

All of the great names were engaged in *philosophy*, not in some arcane speciality.

2   What is the philosophy of mathematics anyway?

Chapter 2 began as naïvely as possible, consulting dictionaries as a stab at answering a 'what is mathematics?' question. For a second question, not about mathematics but 'philosophy of', let us consult the two most recent and most reliable encyclopedias of philosophy published in English. I do not do so because they are definitive, but because a lot of knowledgeable effort and reflection goes into organizing encyclopedias – although perhaps not as much as goes into dictionaries.

There is the online *Stanford Encyclopedia of Philosophy*, and there is the *Routledge Encyclopedia of Philosophy*, henceforth *Stanford* and *Routledge*. After consulting them both, an innocent abroad might be left wondering what counts as the philosophy of mathematics itself. *Stanford* has a heading, 'Philosophy of mathematics' (Horsten 2007). *Routledge* does not. Instead it has 'Mathematics, foundations of', which covers much the same waterfront (Detlefsen 1998). That article is not only about what cognoscenti would call Foundations of Mathematics, as represented, say, by the excellent and very active online *FOM*: 'a closed, moderated, email list for discussing *Foundations of Mathematics*'. It is about the philosophy of mathematics understood in a more general way.

Many topics are discussed in both encyclopedia entries. Yet although between them they list some 170 items in their bibliographies of classic contributions, only thirteen of these occur in both lists. Only eight cited by *Stanford* were published after *Routledge* appeared, so that does not explain the discrepancy.

A complete accounting of all related entries in the two encyclopedias as a whole generates more overlap. Thus *Stanford* has 'Naturalism in the philosophy of mathematics' and 'Indispensability arguments in the philosophy of mathematics', and *Routledge* has 'Realism in the philosophy of mathematics'. Nevertheless, the initial contrast is striking. It is obvious that the general editors of each series had different organizing ideas. Thus *Routledge* does have an entry, 'Philosophy of physics', directing us to more specialist entries, and an entry 'Philosophy of economics', but none for 'Philosophy of mathematics'. That is consistent with the editors having my view that the perennial topic 'philosophy of mathematics' is *not* a specialist topic, but a matter of general philosophy.

At any rate, the prudent innocent, on first looking into these encyclopedias, may judge that the philosophy of mathematics covers a lot of subjects, and that different philosophers have different opinions of what is most central or interesting. The two essays are invaluable introductions to recent work: any reader unfamiliar with what has been going on is well advised to study them. The present book deliberately ignores much of the contemporary material that they report.

3 Kant: in or out?

Neither Detlefsen nor Horsten, neither *Routledge* nor *Stanford*, was addressing my question of *why* there is philosophy of mathematics. Why should they? It was not on the dance card. Both authors report philosophical reflections on results that have arisen in mathematics as a special science, comparable to the philosophy of physics. But a paragraph that might surprise many analytic philosophers and logicians comes right at the start of Detlefsen's survey in *Routledge*. It follows his assertion that Greek and mediaeval thinkers 'continue to influence foundational thinking to the present day':

> During the nineteenth and twentieth centuries, however, the most influential ideas [in the foundations of mathematics] have been those of Kant. In one way or another and to a greater or lesser extent, the main currents of foundational thinking during this period – the most active and fertile period in the entire history of the subject – are nearly all attempts to reconcile Kant's foundational ideas with various later developments in mathematics and logic. (Detlefsen 1998, 181)

Kant does not loom so large in most other introductions to the subject. He is not even *mentioned* in Horsten's *Stanford* 'Philosophy of mathematics'. But this very difference gives us a clue. Kant provides at least one important answer to my question of why there is philosophy of mathematics (at all). I call it an *Enlightenment* answer, Kant being the star and culmination of that epoch in European intellectual (and political) history.

Then, as Detlefsen says, there are the Greek and mediaeval thinkers. Before the Enlightenment answer to my question, there is what I shall call an *Ancient* answer. The two answers focus on different aspects of mathematics.

But hold! I said Kant provides 'at least one important answer'. I shall ignore what many other thinkers, with a far deeper comprehension of Kant than I, would say is the most important influence of Kant upon philosophical thinking about mathematics. It is perhaps best captured by his word, 'intuition'. Modern English usage of that word is very different from Kant's. As announced in §1.30, I shall not use the word in any of its many senses. It is clear that Kant employed his notion of intuition to structure a profound way of thinking about mathematical judgements. I am in no way qualified to discuss that. My Kant is a superficial one, for which I apologize.

Some would say that my use of Kant is worse than superficial: it is Russellian. That will become all too apparent in §23 below. That is a guarantee of bad history, but not always a sign of bad philosophizing. I say a few more words about this at the end of the chapter, §29.

There is indeed a quite different way of answering the question of why the philosophy of mathematics is perennial. *The infinite*. To give that answer would be to write an entirely different book. One might start with the discovery of incommensurability at the time of Pythagoras, but one could find many another more speculative origin. We would traverse Berkeley's critique of Newton's infinitesimals. Kant would enter with the idea of number deriving from succession in time. In the mind of those whom we identify as 'intuitionists' that suggests that number is merely potential. The notion of a totality of integers is superimposed by our understanding on our intuition. (Recall our mention of Bernays in §1.10, to be developed in §§7.2–7). The items I mention in §1 above, such as the continuum hypothesis, would then become central to the story rather than, as in my treatment, peripheral.

To complete this recitation of apologies, my use of Kant completely ignores his reflections on the philosophy of nature, as expounded, for example in his *Metaphysical Foundations of Natural Science* of 1786. This aspect of Kant's thought has been brilliantly analysed in Michael Friedman's *Kant and the Exact Sciences* (1992). It also ignores those who feel that Kant did not know much about even the mathematics of his own day; few, however, are as contemptuous as Cantor writing to Russell, 19 September 1911: 'yonder *sophistical philistine*, who was *so bad a mathematician*' (Russell 1967: 227). Russell remarks that after reading the letter no one will be surprised that Cantor spent a large part of his life in a lunatic asylum – but spent his lucid intervals inventing the theory of transfinite numbers. Russell thought him to be 'one of the greatest intellects of the nineteenth century' (1967: 226).

4   Ancient and Enlightenment

I suggest, then, that the reasons why there is perennial philosophy of mathematics divide roughly in two, Ancient and Enlightenment. The emblem of the Ancient, here, is one strand in Plato. The emblem of the Enlightenment, here, is one strand in Kant's thought. The two strands interweave promiscuously, but it is helpful to distinguish them. (The thoughts of Plato and Kant are so much larger than the strands that I am singling out! Shame!)

The Enlightenment strand is more impressive to philosophers than to mathematicians. The Ancient strand is the one that appears most strongly when mathematicians express philosophical views. The issues in both cases have been discussed intermittently but, it seems, eternally. Hence I shall say nothing new about those famous issues, but the classification proposed here may rearrange some of their relationships, and hold them up to the light in novel ways.

Another caveat. I am concerned with why there is philosophy of mathematics at all. That is a focus for this entire book. Any serious philosophical reflection on maths must surely involve issues that arose in mathematical logic, set theory, and the foundations of mathematics during the twentieth century. I apologize: whatever my opinions and prejudices – and they are unavoidable – they would add little to current discussions. They would just be another item in ongoing discussion. I have a slightly different project in hand.

Like Chapter 1, this chapter will be divided into two relatively independent parts: *Ancient* and *Enlightenment*. This division parallels that of Chapter 1, but in reverse order. A central aspect of the Ancient answer to the why question is none other than Proof, (B) of Chapter 1. A central aspect of the Enlightenment answer is *one* notion of applying mathematics, namely Kant's. That's Application, (A) of Chapter 1. Philosophizing about mathematics comes back to proof and use, over and over again.

A   AN ANSWER FROM THE ANCIENTS: PROOF AND EXPLORATION

5   The perennial philosophical obsession . . .

Why has mathematics mattered to so many famous dead philosophers, from Plato through Wittgenstein? And why does it infect, in many cases, their entire philosophies? Aside from the nay-sayers such as Mill, it is first of all because they have *experienced* mathematics and found it passing strange. The mathematics that they have encountered has *felt* different from other more familiar experiences of observing, learning, discovering, and using new knowledge. This is partly because the gold standard for finding out in mathematics has been demonstrative proof. Not any old proof, for boring things can be proved in boring ways. I mean proofs that deploy new ideas in unexpected ways, proofs that can be understood, proofs that embody ideas that are pregnant with further developments. Mathematicians still like to cite Euclid's proof that there are infinitely many primes, or the proof that the square root of 2 is not a rational fraction. It is, to use the terminology of Chapter 1, the experience of *cartesian* proofs that has attracted many famous Western philosophers to mathematics. Or what Wittgenstein called 'surveyable'. That is why I put the proof of the theorem about not dissecting cubes up front, in my fourth epigraph. How better to convey this experience of finding out something new, 'just by thinking', than by giving an example?

This experience of finding out by thinking, of being certain beyond ordinary certainty, suggests the image of exploration. Alas, this gets hyped into the exploration of something sublime, wholly independent of the mundane realm of feeble humans, and yet discoverable by hard work and genius. But we do not need fantasy to fuel the obsession. There is this feeling that what is being investigated is as much 'out there' as the results of analytic chemistry, or, to use the more familiar metaphor, the upper reaches of the Amazon or the 'canals' of Mars that have been or will be observed and probed by explorers. We might call it the 'stout Cortez' motif of §1.27; to quote more Keats, the mathematician feels 'like some watcher of the skies / When a new planet swims into his ken'. (An ironist will suggest that it makes the metaphor more rich that it was not actually Cortez 'on a peak in Darien' who saw the Pacific, but Balboa, and the last time a new planet swam into a watcher's eyes was Uranus, observed by Herschel in 1781. Keats' poem dates from 1816.)

Let us not forget the warning in §1.22. What I call cartesian proofs engender a particular type of experience, but most proofs are not like that. They can at best be checked, in a leibnizian way. Some may hold that mathematics needs the really good proofs, the ones that really can be grasped as a whole and which bring understanding. Or one may move to Grothendieck's extreme (§1.26), and say that mathematics aims at making the subject matter and its theorems 'obvious', that is, *evident*. More cautiously, we urge only that proofs with a curious quality are what generate the Ancient strand in philosophy of mathematics. It is no accident that popular expositions repeat the same best examples of proof (and why I went out of my way to begin with the cube theorem, familiar to hardly anyone).

6   The perennial philosophical obsession . . . is totally anomalous

Most people do *not* respond to proofs of a certain kind with such experiences, feelings, or fantasy; they really have no idea what is moving those philosophers. They are in good company. Take Hume, one of my heroes who can do no wrong. He was one of the most brilliant reasoners who trod the face of the earth, but it is quite possible that he was never especially impressed by deductive proofs or other types of mathematical argument. 'Experiencing mathematics' in no way implies the possession of philosophical gifts. Perhaps the opposite. But, to vary a letter in T. S. Eliot's *Four Quartets*: 'Hume is where one starts from. As we grow older / The world becomes stranger, the pattern more complicated'. (In *East Coker* v; the initial noun is *Home*, not *Hume*.)

One strange thing about the philosophers' fascination with mathematics is that most people pay it no heed at all. For the majority, like Fuka-Eri of §2.5 above, mathematics is something hateful that is imposed by teachers, and is to be escaped as quickly as possible. Or it is something we are glad other people can use in order to build bridges (an instance used suspiciously often) or to build models to beat the stock market (or usually not).

Few intellectuals are as honest in print as Bruno Latour (2008: 443f, to whom we shall return in §4.13): 'If you have had to suffer through geometrical demonstrations at school (which is certainly the case for me)'. Sounds like he hates maths as much as most mortals – though in fact I am quoting from his brilliant review of Reviel Netz's book (1999) on Greek geometry.

Today's cognitive science is very plausible when it maintains that there are 'innate' cognitive modules, or core capacities (Carey 2009) of arithmetic and geometry that have evolved, perhaps around the time that linguistic modules evolved. But this must be put beside the fact that whereas a high degree of linguistic competence is universally obtained early in life, even modest mathematical competence beyond rote learning is acquired by only a small proportion of the population. To emphasize this contrast, recall Chomsky's insistence on the creative aspect of language use. He claimed that this is a human universal. Then notice that the capacity for even modestly creative uses of mathematics is not widely shared among humans, despite our forcing the basics on the young. So the philosophical obsession is all the more puzzling.

Chomsky taught that language is 'species-specific' – a capacity peculiar to our species. Mathematics is equally species-specific, even if some preliminaries, some sort of number sense for example, are shared with many other species such as crows, dolphins, and macaque monkeys. Yet although most members of our species have some capacity for geometrical and numerical concepts, very few human beings have much capacity for doing or even understanding mathematics. This is often held to be the consequence of bad education, but although education can surely help, there is no evidence that the vast disparity in talent for, or even interest in, mathematics, is a result of bad pedagogy.

A paradox: we are the mathematical animal. A few of us have made astonishing mathematical discoveries; a few more of us can understand them. Mathematical applications have turned out to be a key to unlock and discipline nature to conform to some of our wishes. But the subject repels most human beings.

After that reality check, let us continue with the fact that mathematics has intrigued some and only some philosophers from the time of ancient Greece.

7   Food for thought (*Matière à penser*)

To speak of an Ancient strand in thinking about mathematics is not to be stuck in the past. We are discussing perennial concerns. So I shall use a friendly debate in which a mathematician, who thinks that mathematics is just 'out there', faced off against a neurobiologist who holds that mathematics is all in the human brain. The two protagonists were Alain Connes and Jean-Pierre Changeux. Their discussion was published in French in 1989 as *Matière à penser*, and in 1995 in English translation, with an additional discussion.

Connes, the mathematician (Fields Medal 1982), won another award in 2000 'for revolutionizing the field of operator algebras, for inventing modern non-commutative geometry, and for discovering that these ideas appear everywhere, including the foundations of theoretical physics' – another instance of unity underlying diversity (§1.12 above). He was chosen to announce the seven Millennium Prizes in Paris, 2000, funded by a Boston millionaire, and intended to replace Hilbert's problems announced in Paris in 1900 (§2.21). Changeux is an eminent neurobiologist. The two men are colleagues at the Collège de France in Paris.

Connes cannot doubt that there is a mathematical reality, independent of human thought. 'The working mathematician can be likened to an explorer who sets out to discover the world' (1995: 12). This is the exact same metaphor used by G. H. Hardy in his well-known essay, 'Mathematical proof' (1929). Notice that Hardy, despite his title, emphasized exploration rather than formal proof. There is the oft-quoted line, 'proofs are what Littlewood and I call gas':

> there is, strictly, no such thing as mathematical proof; that we can, in the last analysis, do nothing but *point*; that proofs are what Littlewood and I call *gas*, rhetorical flourishes designed to affect psychology, pictures on the board in the lecture, devices to stimulate the imagination of pupils. (Hardy 1929: 18)

For Hardy, it is not the experience of proving that creates his sense of mathematics, but the experience of coming to see, by proof or any other kind of activity that produces insight. I want to emphasize the *experience* that leads one to feel like 'stout Cortez'.

Some fifty-five years later, Michael Atiyah (first mentioned in §1.12) expressed a related thought in a much reproduced interview: 'I may think that I understand, but the proof is the check that I have understood, that's all. It is the last stage in the operation – an ultimate check – but it isn't the primary thing at all' (Atiyah 1984: 13). We can update this to 2011, as reported in §1.22: Voevodsky's proposal that every published proof will soon come in machine-readable form, with a proof-checker verification. That would truly be the last stage in the operation. No longer a rhetorical flourish intended to affect psychology (for it does not affect psychology at all, in the sense that Hardy meant).

8 The Monster

The 'Ancient' source of philosophy of mathematics lies in the experience of proof and, more generally, of mathematical discovery, and, finally, the sense that mathematics is just 'out there'. Connes elegantly illustrates the experience of something strange and profoundly impressive about the mathematical life. It illustrates why there is a philosophy of mathematics at all. Connes is enormously impressed by a fact about mathematical research. His example is recent, yet is of just the same form as examples that could have been used in classical Athens.

> Here we come upon a characteristic peculiar to mathematics that is very difficult to explain. Often it's possible, although only after considerable effort, to compile a list of mathematical objects defined by very simple conditions. Intuitively one believes that the list is complete, and searches for a general proof of its exhaustiveness. New objects are frequently discovered in just this way, as a result of trying to show that the list is exhausted.

Take the example of finite groups. The notion of a finite group is elementary, almost on the same level as that of an integer. A finite group is the group of symmetries of a finite object. Mathematicians have struggled to classify the finite simple groups, that is to say the finite groups that (like the prime numbers to some extent) can't be decomposed into smaller groups ... Fifteen years ago [namely 1974] the last finite simple group – the 'Monster' – was discovered by purely mathematical reasoning. It is a finite group with a considerable number of elements:

808,017,424,794,512,875,886,459,904,961,710,757,005,754,368,000,000,000

It has now at last been shown, as a result of heroic efforts, that the list of twenty-six finite simple sporadic groups is indeed complete. (Connes, in Changeux and Connes 1989, translated in 1995: 19f. I have inserted the commas in the size of the Monster.)

Connes writes as if the Monster has been sitting out there, quietly grinning, waiting for us to discover it.

9   Exhaustive classification

Connes concludes the above quotation by saying that all the different types of finite simple groups have been enumerated – as the result of heroic efforts. There are twenty-six of them. This result is called a classification theorem. To continue the analogy with exploration, some may feel that it is a bit like enumerating all of the 4,000 metre peaks in Europe (that's eighty-two official summits plus forty-six lesser summits). It is as if all the different types of finite simple groups were just 'out there' waiting to be located, just like the peaks.

Classification theorems play a major role in modern mathematics. The complete, exhaustive classification of the finite simple groups was, as Connes reports, known by the end of the 1970s. What is regarded as its definitive proof was published only in 2004 (always subject to the proviso that there still *might* be a counter-example lurking in the wings, 'out there').

The first classification theorem in the history of the human race is Euclid's proof, in Book xiii of the *Elements*, that there are exactly five regular polyhedra. He first shows how to construct them, and then proves that is all there are. It has been proposed that the entire point of the *Elements* was to reach this very peak of discovery.

I should warn that although my example from Connes is a classification theorem, and although I shall often refer to Euclid's result, classification theorems are just one type of mathematical discovery. It is true, however, that they especially lend themselves to metaphors of exploring a terrain 'out there'.

10 Moonshine

Mathematicians soon made some sense of the Monster. The Monster might possibly be identical to an object derived from a completely different branch of mathematics, the theory of elliptic functions. It *had* to be a coincidence! The first reaction of the prolifically creative mathematician John Conway – who had named the Monster in the first place – was 'Moonshine!' (referring not to the stuff distilled in the Appalachian Mountains but to the effect of phantasms on water in the light of the full moon). So that idea was called the *Monstrous Moonshine conjecture*. Except it was instead another of those cases of underlying unity within the diversity of mathematics.

That was in 1973–5, a period of intense activity whose temporary epicentre was Cambridge, UK, but with much input from, for example, Michigan and Bielefeld. (When Connes is translated above as saying 'fifteen years ago', that is translating, without alteration, the French text of 1989; hence my insert of 1974.) Robert Griess (1982) provided the first construction of the Monster, which he called the Friendly Giant; that paved the way to checking out the moonshine.

For one of several popular expositions of this story, see Ronan (2006). If you fancy tales of 'mad mathematical genius', read the account of Conway's close collaborator Simon Norton (Masters 2011). Masters happily declined to speculate about the autism spectrum (§2.9).

> 'Everyone thinks about that the instant they see him', Masters says. But the author didn't want to bring this question into the book, and readers won't find the word 'autism' anywhere in it. 'It seems to me, you take Simon as Simon is', Masters explains. 'If you're born with this degree of genius, this degree of capacity, you're going to grow up a bit weird.' (Masters 2012)

More importantly Simon Norton exemplifies the importance of *play* to a certain type of mathematical mind (§2.27). Even the riddler Dudeny, quoted in that section, shows up in Masters' playful book (2011: 123). And for Conway on serious mathematics + play, see ONAG, the common name for *On Numbers and Games* (Conway 2001).

11 The longest proof by hand

In 1989, Connes, like many others, had no doubts about the 'facts', that is, the existence and size of the Monster and, more importantly, the completeness of the enumeration of the simple finite groups. Not everyone agreed. As late as an interview for his Abel Prize, 2 June 2003, another eminent

French mathematician, Jean-Pierre Serre (2004), was asked if he still had doubts about the extent to which this fact had been proven. (He had been expressing reservations during the 1990s.) He replied that he understood that a proof was just then probably about to be completed.

Indeed it was. A year later Michael Aschbacher and Stephen Smith (2004) published a proof in two volumes that no one person will ever, in the future, read. That is perhaps the most compelling example of a non-cartesian proof.

Ronald Solomon (2005) said in a long review for the American Mathematical Society that no single person would read the second volume, but a team had checked it. He emphasized that everything in the published two-volume proof was done by hand. It was *not* a computer-generated proof, although computers were much used in exploring the terrain.

12   The experience of out-thereness

Here I want only to emphasize Connes' heartfelt *feeling* or *experience* that this at first sight absurd object was *just there*, waiting for us. And a little later, this monstrous object turned out to be the very same as one identified in a completely different field of mathematics. Out there? We are reminded of Frege's example: the Morning Star turned out to be identical to the Evening Star. Thus comes the idea of an object, a mathematical object, waiting to be identified, like Venus.

Connes' reaction is quite a common one. Richard Borcherds won the Fields Medal in 1998 for, among other things, proving the Moonshine conjecture. In conversation he said that, 'When you think about the Monster you have to wonder who *made* it! It is almost like that Intelligent Design stuff; the Monster has such a complex and yet organized structure, that it is as if it had been engineered by someone' (quoted by permission from a conversation on 27 July 2010).

Borcherds was honestly expressing his persisting astonishment about the sheer existence of the Monster. That kind of experience serves to cultivate the Ancient strand in the philosophy of mathematics, and is quite possibly where it all began.

By the way, with respect to mathematicians and Asperger's syndrome (§2.9), Simon Baron-Cohen (2003: ch. 11) dedicated a whole chapter to Borcherds and his relatives. He was interested in how autistic traits run in the male side of families; Borcherds' father is a physicist, and his brothers are mathematicians, so the group is a promising case study. Baron-Cohen concluded that such a diagnosis was not justified, but a close thing.

13   Parables

Borcherds does not imagine that the Monster had a designer; he said only that the object is so delicate in all its parts that it is as if it had been designed. He was not advancing an opinion, let alone an argument, for 'intelligent design'. He was expressing heartfelt incredulity at a fact that, in any ordinary sense, he understood at least as well as anyone else in the world. He was not so much surprised at the fact as by the *existence* of such facts.

An apocryphal story is told of many very smart old non-mathematicians. I first heard it as told about Hobbes. There is no reason to believe it is true of *anyone*, but it is a great parable. Hobbes goes into a noble library and sees a book open on the table. It is Euclid. Hobbes reads the proposition announced on the page before him. 'Absurd', he says. 'Can't possibly be true.' He goes back to the axioms and definitions and reads the book through. He is astonished. Yes, it is true.

That old chestnut is an exact mirror of recent history. Conway, not a man given to mincing words, exclaims 'Moonshine!' That's *exactly* what 'Hobbes' might have said (although in his day the expression was 'moonshine on water'). Finally an argument, a proof, or proof sketch, convinces the sceptic that the moonshine is the exact truth.

We can imagine this story having been enacted in ancient times. One of Plato's pupils comes across the assertion that there are exactly five regular solids. Moonshine! Then he dutifully reads through Euclid, right up to the end. He is convinced.

14   Glitter

Connes was giving vent to exactly the same sentiment as the apocryphal Hobbes and my fictional pupil of Plato. One may complain, however, that he was also using a certain sort of mathematician's rhetoric to scare us into submission. We don't understand all this stuff about the Monster, but Wow! Ain't that something!

We are reminded of Wittgenstein's talk of *glitter*. The metaphor can be applied to Connes' rhetoric. We are supposed to be impressed when he writes down this meaningless (to us) fifty-four-digit number at the end of my quotation in §8 above. Somehow it is overdone; instead we're offered what Wittgenstein called '*mysteriousness*' (*RFM*: v, §16d, 274, original emphasis).

'All that I can do [Wittgenstein continued] is to shew an easy escape from this obscurity and this glitter of the concepts' (§16e). He was thinking about Cantor's theory of transfinite cardinals. All too easy an escape, cynics will complain; theft, or rather rubbishing, rather than honest toil. But those infinities are perhaps not inevitable; we will discuss Paolo Mancosu's questioning of that in §4.3. If you think you need transfinites to make the calculus rigorous, help yourself to hard work, namely Abraham Robinson's non-standard analysis. But there is no comparable escape from the 'glitter of the concepts' illustrated by Connes. The Monster, once moonshine, is almost certainly a fact. It just *is* astonishing, that careful, long, and collective reflection on so elementary a concept of finite simple groups should generate this strange 'Monster', even if by now it is well enough understood to call it a Friendly Giant.

15   The neurobiological retort

Jean-Pierre Changeux was not impressed by the glittering Monster. He holds that mathematical truth is constrained by the neuronal structure of the brain. He has no use for Connes' conception of a mathematical reality 'out there'. But he does not doubt that mathematical objects exist. He is a neurobiologist with access to the resources of brain science and the cognitive sciences. He is convinced that mathematical structures are by-products of the innate endowments of the human brain. He goes so far as to say that *'mathematical objects exist materially in your brain'* (Changeux and Connes 1995: 13, my italics). Neurobiological realism about mathematical entities!

Responding to Connes on the Monster, he retorts that we have here only a complicated version of the finite list of regular polyhedra, a list that so impressed Plato that they are called the Platonic solids.

Changeux's taunt, 'Same old story', may backfire, but not with ill effect. I imagined Plato having used the regular polyhedra in an argument identical to that of Alain Connes, but with the classification of the regular polyhedra in place of the classification of the finite simple groups. Indeed a popular exposition of the Monster and its confrères by Mark Ronan (2006) actually begins with the Platonic solids. Connes chose something new and glittering precisely because we have become blasé about polyhedra, but the point is exactly the same. (And indeed I chose the cube theorem because we have become blasé about the proof of the irrationality of √2.)

Of course what Changeux is really protesting is glitter: Connes is doing a snow job. The Platonic solids are merely last year's glitter. If you're not impressed by them, don't be bowled over by the Monster. Most mathematicians will retort that we do continue to be bowled over by the Platonic solids. Michael Atiyah and a colleague wrote a marvellous survey paper, 'Polyhedra in physics, chemistry and geometry', that begins by enthusing about the self-same Platonic solids; cf. §4.10 below (Atiyah and Sutcliffe 2003).

16 My own attitude

I have often, rightly, been accused of laying out a set of options quite well, but concealing my own hand. What on earth do *I* think? Should this be taken as a criticism? I could well plead some precedents, including Socrates and Wittgenstein. But that would be hubris, and I shall not take that route.

What do I think about Changeux versus Connes? It really does not matter much what I think. The debate between Connes and Changeux wears contemporary garb, but its form is perennial. It has a more conventional presentation in another conversation, Connes *et al.* (2000), this time including André Lichnerowicz, a structuralist mathematician, in the sense of Bourbaki §2.12. We shall turn to that conversation in §5.7–13. But neither I nor anyone else is going to settle anything. My personal opinion is as worthless as almost everyone else's. It would, however, be disingenuous not to declare my sympathies. They are, unfortunately, rather more obscure than that of either of these two controversialists.

I think that what Connes *says* is right but what he *means* is wrong. He really does believe in a reality out there, and, as we shall see in Chapter 6, he also believes in, for example, the series of whole numbers, archaic and primordial, whose structures have nothing to do with the brain, except that our brains enable us to apprehend them. That (I assert) is what he *means*, and in my opinion that is wrong.

I think that what Changeux *says* is wrong – 'mathematical objects exist materially in your brain' – but what he *means* is right. Neurons exist in the brain, blood exists in the brain, flesh exists in the brain, but numbers don't. The Monster does not exist in the brain. Some current work talks of specific neurons coding for numbers. Maybe that will pan out, but it is a very long way from saying that numbers exist in the brain. And it is a very long-term research programme to identify 'flesh-and-blood' structures in the brain as somehow 'representing' numbers. But it is a worthy programme.

I believe that what Changeux *means* is that the objects that Connes so admires are by-products of human mental activity, which is determined by our genetic envelope (a phrase I got from Changeux himself, but which he seems not to use in print). I believe that is right, but that we do not understand it at all well, and that it is only part of a story.

Mathematics is a human activity, grounded in the body, hands as well as brain, and formed by human communities at very specific times and places.

Changeux speaks to one part of that: neurological states of affairs that are preconditions for mathematical thought. They are part of what I call a cognitive plateau of capacities that humans have discovered how to use in order to engage in mathematical thinking. But he does not even address how that happens. In my opinion we have to move on from that plateau by engaging in what I shall call cognitive history. That is a term I adapt from Reviel Netz (1999).

17   Naturalism

Some readers, perhaps most, will want to catalogue what I have just said under an increasingly prominent body of philosophical writing about mathematics, which is called *naturalism*. It owes much to Quine's naturalized epistemology. Since I don't think of my work in that tradition, I do not file it that way myself.

'Epistemology, or something like it, [Quine wrote] simply falls into place as a chapter of psychology and hence of natural science' (1969: 82). Quine has generated a picture of epistemology as a search for foundations, which is to be replaced by a study of the reasons people actually give, a topic for psychology. Foundations for knowledge, however, have never been of the slightest interest to me – thanks, perhaps, to reading Peirce and Popper.

I do believe that in order to understand how mathematics became possible for a species like ours, in a world like this one, we need to draw on a great many kinds of inquiry, some of which fall under psychology and some of which are natural sciences. These include evolutionary biology, cognitive sciences, developmental psychology, and neuroscience, but also archaeology (especially the archaeology of mind or cognitive archaeology), prehistory, anthropology, ecology, linguistics, sociology, science studies, mathematics itself, and good old-fashioned history, among others. Not to mention philosophy of various stripes. Probably phenomenology has offered more in the way of discussion of these questions than analytic philosophy.

One outstanding self-avowed proponent of mathematical naturalism is Penelope Maddy (1997, 1998, 2001, 2005, 2007), whose ideas have much evolved from 1997 to 2007. One of the leading ideas remains, however. 'Mathematical naturalism' is a generalization of a view about method, 'namely, that mathematical methodology is properly assessed and evaluated, defended or criticized, on mathematical, not philosophical (or any other extra-mathematical) grounds' (Maddy 1998: 164).

Of course! Who could disagree? Even Berkeley criticizing Newton, though motivated by his philosophy, was arguing on what then counted as mathematical grounds.

I hope it is also clear that I have no interest *whatsoever* in assessing, evaluating, defending, or criticizing mathematical methodology. Maddy does philosophers a service by reporting some elements of recent developmental psychology cum cognitive science (2007: ch. III.5). She leads us up to what I call the cognitive plateau of core capacities that make arithmetic and geometry possible. In this and many other respects I share many interests with serious naturalists. If that makes me a naturalist, so be it.

An early exponent of mathematical naturalism was Philip Kitcher (1983). When I first read his ideas, they sounded like an update of John Stuart Mill's thesis, in his *System of Logic*, that the truths of mathematics are the most general truths of experience (Hacking 1984).* I do *not* believe that mathematical truths are most general truths of experience, although I very much respect what Mill was saying, and why he was saying it. A bit more on Mill at the end of this chapter, §34.

You can, in fact, see Quine's holism as going in the same direction, except organizing 'experience' in terms not just of generality but of centrality and hence of resistance to revision.

I resist the label 'naturalist' because my way of thinking does not come from that Quinean tradition. It comes from the world in which we all live. If we want to understand why there is philosophy of mathematics, we have to think about the phenomena (natural? evolutionary? human? historical?) that bring it into being. Undoubtedly that aligns me with naturalism. Yet it is formally compatible with the boldest type of pythagoreanism-cum-platonism, of the sort advocated by Roger Penrose (1997). The difference is only that a pythagorean naturalist has somewhat different natural phenomena to explain. I suggest in §6.13 that pythagoreanism might turn out to be the deepest naturalism.

After these quibbles, I should say that although I find 'naturalism' to be a rather shallow label, I learn much from those who profess it; and also from those who try to refute it, such as Brown (2012). My resistance to the label is itself of no importance. In §6.14 we turn to a philosophical discussion by the mathematician Tim Gowers, whom we have come across already in §2.18. Early on in his talk he said, 'I take the view, which I learnt recently goes under the name of *naturalism*, that a proper philosophical account of mathematics should be grounded in the actual practice of mathematicians.' In fact, I should confess that I am a fan of the later Wittgenstein' (Gowers 2006: 184). If *that* is what naturalism is, count me in. But few philosophers who self-identify as naturalists appear to have much use for the later Wittgenstein.

18 Plato!

One reason, the ancient reason, why there is philosophy of mathematics at all is that many mathematicians have exactly the same inclinations as Alain Connes. This is because they experience certain phenomena, of proof and exploration and solution, very intensely. They cannot resist the family of 'out there' metaphors that come so forcibly to the fore. At the same time, most thoughtful people are like Changeux, and find that story simply unintelligible. But it is not only avowed neuroscientific materialists (reductionists?) who see things that way. I have already quoted Langlands' discomfort. Speaking of the intimation that mathematics exists independently of us, he wrote that 'This is a notion that is hard to credit, but hard for a professional mathematician to do without' (Langlands 2010: 6).

That's what's called a philosophical problem, or a thesis and an antithesis. Hegel taught that we should overcome both ways of thinking, resolving the problem in a 'synthesis'. There is still no agreement on what the synthesis ought to be. People who care seem to cling frantically either to the thesis or the antithesis, although of course a great many people who are interested don't really care.

It is no accident that in trying to see why there is philosophy of mathematics at all, the name of Plato should at once have loomed up large. Plato and the Pythagoreans are the first recognizable philosophers of mathematics. Ever since, we have been chewing over bones similar to those on which they cut their teeth.

I emphasize that I have been discussing an *ancient* reason for the very existence of the philosophy of mathematics. I have used a contemporary mathematician, Alain Connes, to illustrate that this is still a powerful reason. It has very little to do with the philosophy of mathematics now called platonism. That is a doctrine about the existence of mathematical objects, and more generally of abstract objects. We owe the label 'platonism' to the logician Paul Bernays (1888–1977) writing in 1935. He wanted a name to contrast with the triad of competing schools, intuitionism, formalism, and logicism. It has since been turned into the name of a semantic doctrine about reference of names for mathematical objects, starting with the numbers. We shall discuss these matters in Chapter 7B.

I do not say that platonism has *nothing* to do with Plato! Only that the way these issues are now posed requires a twentieth-century background. In contrast, what has been described in this chapter thus far, the experience of discovery, the sense of what is found being 'out there', is one thing that moved Plato just as it moves Connes. It requires nothing more than, shall we say, a fourth-century (BCE) background. Hence I call Connes a capital *P* Platonist. As already announced in §1.18, I use an orthographic convention: capital *P* Platonists for those whose motivations are primarily ancient, and lower-case *p* platonists for those whose analyses are primarily grounded in twentieth-century semantics. Probably most platonists are also Platonists at heart, and many think that semantic platonism is the right way to express Platonist instincts. These matters are postponed to Chapter 6.

B AN ANSWER FROM THE ENLIGHTENMENT: APPLICATION

19 Kant shouts

Recall Kant's glorious exclamation of awe, the one that created a philosophical language that is still in use. The *Prolegomena to any future Metaphysics that will be Able to Come Forward as Science* (1783) comes after the first edition of *The Critique of Pure Reason* (1781).

> Here now is a great and proven body of cognition, which is already of admirable extent and promises unlimited expansion in the future, which carries with it thoroughly apodictic certainty (i.e. absolute necessity), hence rests on no grounds of experience, and so is a pure product of reason, but beyond this is thoroughly synthetic. 'How is it possible then for human reason to achieve such knowledge wholly *a priori*?' (Kant 1997: 32)

In the preface to the second edition of the first *Critique* (1787), he condensed the final question into the more memorable but less clear, 'How is pure mathematics possible?' There is our philosopher's vocabulary for discussing mathematics:

- necessary/contingent
- analytic/synthetic
- certain
- a priori.

Kant did not just list the problematic ideas. He shouted. Not just certainty, but *apodictic* certainty! *Absolute* necessity! *Thoroughly* synthetic! *Wholly* a priori. We may call these *hyper-adjectives*.

Kant's focus is rather different from the ancient – and still present – concern that there is a body of mathematical facts just 'out there', independent of the material world that we inhabit and apparently independent of the human mind.

I shall say nothing here about apodictic certainty, but will briefly begin with absolute necessity. Most of the rest of this chapter will be about the synthetic a priori, although I shall try to avoid the phrase, which has ceased to be helpful except as a memento of the past. First a word on terminology.

20   The jargon

Kripke (1980) did yeoman service in making readers aware that it was foolish to equate the labels 'necessary', 'a priori', and (before Quine) 'analytic'. To treat them as more or less equivalent is worse than mere carelessness. It is a category mistake, confusing attributes of knowledge and attributes of propositions. 'A priori' is a predicate of knowledge, or better, as in Kant, of judgement. So is certainty. 'Necessary', as we now use the term, is a predicate of propositions, as are 'analytic' and 'synthetic'. These confusions were exacerbated, if not created, by the logical positivist account of logic and mathematics to which we turn briefly in §29 below.

In my own experience, careful analytic philosophers did not fall prey to the confusions that seemed to flow from logical positivist doctrines. That was certainly true of my own teacher, Casimir Lewy,* who drilled the distinctions into his pupils. For my own version of the distinctions among 'analytic', 'necessary', 'a priori', and so forth, that I picked up from Lewy (probably in 1957), see Hacking (2000b: §5).

Kripke's reminder of the topography of these terms was nevertheless an invaluable piece of hygiene for those not so trained. To be grateful for his clarification is in no way to commit to the usefulness of his idea of a posteriori but necessarily true statements of identity. I regard that idea of his as a by-product of a notation for identity, in much the way that, in the *Tractatus*, Wittgenstein may have held that tautologies are by-products of the introduction of the logical constants. But that is not a matter that concerns us here.

The definitive analysis of terminological issues is given, in passing, by Per Martin-Löf (1996 and references at the end). He explains the roots of the historical changes in philosophical jargon that have taken place from Aristotle through Kant and on to the present. He is very good on the dual roles of 'judgement' (*Urteil*) in Kant's writing. We see why all of Kant's labels can serve as attributes of 'judgements'. *He* did not commit a category mistake! There is a good deal to be said for returning to Kant's usage, but I shall not do so systematically.

21 Necessity

You may feel that Kant was giving no more than an Enlightenment expression to Ancient worries. Yes, of course, but importantly no. On the score of necessity, I appeal to Miles Burnyeat's (2000) assertion that Plato had no concept of (logical) necessity. Mathematical truths mattered to him, because they are eternal. Logical necessity, as we know it, is an invention of the modern era, derived from concepts of mediaeval Islam and Christendom, and harking back to very different notions in Aristotle.

Of course it ties in with the Ancient astonishment with proof, for we seem to prove things that *must* be true, that cannot be otherwise. And how can that be? Couldn't an omnipotent creator make things differently? That's why theology matters. It was a question that divided Descartes and Leibniz. Descartes agreed that there are eternal truths, but thought that God could have arranged things differently. Leibniz, in contrast, perfected our modern notion of logical necessity, which he himself helped to create (Hacking 1973).

'It used to be said that God could create anything except what would be contrary to the laws of logic' (Wittgenstein 1922: 3.031). Well, what would stop him? In §1.19 I quoted Wittgenstein's answer, that 'we could not *say* of an "unlogical" world how it would look'. Later, things were less clear. When we prove unexpected consequences – or see on the basis of argument that something could not be otherwise – we become afflicted by the experience of necessity. In one of his characteristic aphorisms – but remember the caveat of §1.1 above – Wittgenstein called this experience 'the hardness of the logical *must*' (*RFM*: 1, §121, 84).

Why is there perennial philosophy of mathematics? Part of the Enlightenment answer lies in the apparent necessity of the laws of logic, and indeed of the whole of mathematics. But the idea that the experience of something having to be true applies to all mathematics is an illusion. It happens at best with facts for which we have cartesian proofs, proofs that convey understanding, proofs that show why the fact proven is true. I have not the slightest inclination to say that the result of a calculation is necessarily true. I pick up a pencil and multiply to obtain 13×23=299. Compare working out a route on the London Underground. I am way out in Cockfosters and want to take my grandchildren to the Tower of London. One of them is rather young, so I'd prefer to change stations only once. Well, one way to do it is to go all the way to South Kensington on the Piccadilly line, and then back to the Tower on the District and Circle. The solution to my problem is *no more and no less* experienced as 'necessary', as something that 'must' be true, than the answer to my arithmetical exercise. It is just what you get by working things out using what you have learned by rote.

22 Russell trashes necessity

The great empiricists of modern times – John Stuart Mill, Bertrand Russell, and W. V. Quine – have done their best to scotch necessity as a philosophical problem. Mill may have been the one who felt most strongly, and that for reasons both political and ideological. He was convinced that the idea of necessity, abetted by the companion neo-Kantian notion of a priori knowledge, is 'the great intellectual support of false doctrines and bad institutions'. His passionate hatred for these ideas is quoted in §34 below.

You might think that Mill's ringing denunciation of Kantian absolute necessity was about as far as you can go, but you can't beat Russell (1903: 454): *The theory of necessity urged by Kant ... appears radically vicious.* (In context, Russell was discussing Lotze's account of space. Rudolf Hermann Lotze (1817–81) is now forgotten, but was taken very seriously not only by the German logicians such as Frege, but also by both Moore and Russell.)

The early bits of Russell's *The Principles of Mathematics* are more dipped into, nowadays, than what comes later (e.g. the just-cited page 454!). So it is quite widely known that Russell carefully distinguished between implication, which he went on to call material implication, and formal implication. Material implication, $p \supset q$, is the relation that holds between two propositions p and q when q is true or p is false. But this is not a (non-circular) definition, for the equivalence here is just two-way implication, which at that time was not good enough for Russell. Hence he asserted that 'implication is indefinable' (1903: 15).* He also introduced the notion of what he called 'formal implication', a relation between propositional functions φ and ψ, such that $(x)(\varphi(x) \supset \psi(x))$.

The relevance to a discussion of necessity is that Russell admits no stronger relation of necessary implication, no entailment, the idea that Moore formulated later, or strict implication, as urged by C. I. Lewis. Russell also thought that philosophers from time immemorial had confused material and formal implication. Thus the once-named eternal truths would typically be universally quantified statements, whose logical form would usually work out as $(x)(\varphi(x) \supset \psi(x))$.

Russell's rediscovery of Leibniz in 1900 set the course of English-language Leibnizian studies from then to now. Leibniz was the arch-patron of necessity, and one might have expected some conversion on Russell's part. Not at all. In 1903 Russell was sure that 'everything is in a sense a mere fact'. 'The only logical meaning of necessity seems to be derived from implication.'

> There seems to be no true proposition of which there is any sense in saying that it might have been false. One might as well say that redness might have been a taste and not a colour. What is true, is true; what is false, is false; and concerning fundamentals there is nothing more to be said. The only logical meaning of necessity seems to be derived from implication. A proposition is more or less necessary according as the class of propositions for which it is a premise is greater or smaller. [Russell here has a footnote to 'Necessity', Moore 1900.] In this sense the propositions of logic have the greatest necessity, and those of geometry have a high degree of necessity. (Russell 1903: 454)

Here he drew on Moore (1900), an essay in *Mind* that tried to sort out several ideas underlying necessity.* Russell concluded his brief discussion by saying that there are at most degrees of necessity, and that in some sense 'the propositions of logic have the greatest necessity, and those of geometry have a high degree of necessity'. That was very much in the spirit of Moore's paper.

Whereas Russell's animadversions on necessity have been forgotten, Quine's continue to be very well known. His *Methods of Logic* (1950) is more than a manual for an introductory course on symbolic logic; it is a philosophical masterpiece in its own right. Its brief Introduction is as good a place as any to go for a succinct statement of Quine's epistemology. The idea of a 'conceptual scheme' is in full bloom. 'Our statements about external reality face the tribunal of sense experience not individually but as a corporate body' (1950: xii). There are the two competing priorities in the struggle for survival against recalcitrant experience. One is proximity to direct experience, while the other is that 'the more fundamental a law is to our conceptual scheme, the less likely we are to choose it for revision' (1950: xiii).

> Our system of statements has such a thick cushion of indeterminacy, in relation to experience, that vast domains of law can easily be held immune to revision on principle. We can always turn to other quarters of the system when revisions are called for by unexpected experiences. Mathematics and logic, central as they are to the conceptual scheme, tend to be accorded such immunity, in view of our conservative preference for revisions which disturb the system least; and *herein, perhaps, lies the 'necessity' which the laws of mathematics and logic are felt to enjoy.* (Quine 1950: xiii, italics added, scare quotes in original)

That pretty much says it all. 'Necessity' enters only when shielded by quotation marks. Whatever necessity pertains to mathematical statements, it is a consequence of their centrality to our conceptual scheme.

In 1950 Quine was well aware of the potential effects of quantum reasoning on our entire conceptual scheme. There could be 'sweeping simplifications' by radical restructuring that infected even the logico-mathematical core. 'Thus the laws of mathematics and logic may, despite all "necessity", be abrogated' (1950: xiv; scare quotes in original). (Incidentally, much of this material was suppressed from the final edition of *Methods*, but not because Quine had changed his mind.)

In matters of necessity as in many other things, Quine and Russell were pretty much birds of a feather, but whereas Russell in 1903 was dogmatic in his ditching of necessity, Quine was holistic. The effect was much the same.

23   Necessity no longer in the portfolio

Artists have portfolios of their work, and stockbrokers have portfolios of stocks. Mary Douglas and Aaron Wildavsky (1982) coined the idea of a risk portfolio. One year we are fearful of a nuclear winter; a decade later we fear global warming. One replaces the other in the public risk portfolio.

Ever since the beginning of analytic philosophy, there has been the sense that philosophers address problems. But there is a portfolio of problems under discussion, a portfolio that evolves over years.

William James' last, unfinished, book was *Some Problems of Philosophy* (1911). G. E. Moore was giving lectures in 1910–11 on 'Some main problems of philosophy' (much later published under that title). And then there was Bertrand Russell's 1912 *Problems of Philosophy*. The problems give a pretty good idea of the 1910 portfolio of philosophical problems being discussed in both the old Cambridge and the new. Many of those problems are in the analytic portfolio (though not the pragmatist one) a century later.

When we become more fine-grained and discuss the philosophy of mathematics, I can say from memory that 'necessity' was rather central to the portfolio in 1960, but it has now pretty much dropped out. Of course logical necessity and modal logic are going strong in their corners of the universe, but worries that some things 'must' be true (and why?), or the feeling of compulsion accompanying a good cartesian proof (what compels?) seem no longer to be what interest students. Quine triumphant! He completed the task that previous empiricists like Mill and Russell had begun.

Even when spurned by philosophers, the feeling of necessity, of logical compulsion, is one of the reasons that there is philosophy of mathematics at all – especially if we include what might be called 'deconstructive' philosophy of mathematics in the style of Mill.

I would deconstruct part of the appeal in a more humdrum way. We just don't have that experience with most mathematics. We falsely generalize from a genuine experience. To relieve the worry of necessity, we should examine the genuine experiences, typically prompted by brisk cartesian proofs, splendidly epitomized by Martin Gardner's (1978) Aha! experience. So much is trivial: it is just the injunction, stop and look! But it does not go anywhere near the heart of the matter.

24 Aside on Wittgenstein

I suspect that one of the reasons that people nowadays have so many problems with the *Remarks on the Foundations of Mathematics* is that some of its central concerns have pretty much dropped out of today's portfolio of problems. I hold that although he hardly used the word 'necessity', related concerns were central to many of Wittgenstein's mathematical concerns in 1938–44 (Hacking 2011b) – and indeed throughout his entire life. Take the book published as *Philosophical Grammar*, written in the early 1930s. Part I, 'The proposition and its sense', is 236 pages long; Part II, 'On logic and mathematics', occupies 247 pages. You might say Part I is philosophy of language, and Part II philosophy of mathematics.

Yet they interconnect essentially. For one of Wittgenstein's thoughts, especially after 1937, was that proof somehow fixes the meanings of the sentence proved. If you like, the proposition *becomes* necessary, is used as a new criterion of what's possible, when it is proved. But this flies in the face of our conviction that the meaning has *not* changed. I say that Wittgenstein was fighting the tyranny of the *Satz*, of the fixed proposition/sentence. He gnawed and gnawed away at surrounding issues, never to his complete satisfaction. One of his last attempts, strictly in the try-out mode ('It is as if', 'It is so to speak') is this:

> It is as if we had hardened the empirical proposition into a rule. (*RFM*: VI, §22b, 324)

> It is so to speak an empirical proposition hardened into a rule. (*RFM*: VI, §23c, 325)

Mark Steiner (2009) lays great weight on the idea of 'hardening'. I am more cautious (Hacking 2011b), as was Wittgenstein at the end of his life, when he was writing the notes published as *On Certainty*:

> Isn't what I am saying: any empirical proposition can be transformed into a postulate – then becomes a norm of description. But I am suspicious even of

this. The sentence is too general ... One almost wants to say 'any experiential proposition can, theoretically, be transformed', but what does 'theoretically' mean here? It sounds all too reminiscent of the *Tractatus*. (Wittgenstein 1969: §321)

The *Tractatus* is the classic work dominated by the *Satz*, the fixed, hard, *thing*, in a universe not based on things, and Wittgenstein was forever fighting it off.

Unlike many other readers, I find that Wittgenstein's discussions of following a rule are very often directed to matters connected with necessity. They are regress arguments, countering the idea that necessity arises from rules. They are curiously analogous to Quine's position in 'Truth by convention' (1936). But this is not the occasion to develop that understanding, precisely because necessity does not, at present, figure much in the portfolio of analytic philosophers who discuss mathematics.

25 Kant's question

The question of 'a priori' knowledge, or, in Kant, a priori judgement, is different from that of necessity. It is double-edged, for it gets its impact from two aspects. One is the thought that we can find out 'just by thinking'. The other is that what we find out is true – true of the world. These together create a terrible conundrum. In the preface to the second edition of the *Critique of Pure Reason*, Kant shouts again. *How is that possible?*

'How is pure mathematics possible?' Kant did not mean exactly what we call 'pure mathematics' as opposed to what we call 'applied mathematics'. As we shall explain at length in Chapter 5A, that terminology arose in the decade in which Kant wrote the first critique, although a similar distinction, between pure and mixed mathematics, had been introduced by Bacon. Very roughly, and somewhat anachronistically, both 'mixed' and 'applied' drew attention to information derived from experience.

I think that nowadays applied mathematics means something more like, 'can be used in the material world'. Thus if I use arithmetic to figure out how many lunches are needed for the primary school picnic, I am applying arithmetic, and 'thus' applying elementary mathematics. Likewise if I use the calculus to compute an annuity I am applying less elementary mathematics. Even the second of these would, I think, count as pure in Kant's eyes. But if I am using the calculus in ballistics or the design of a dam, I incorporate empirical premises, and that is applied.

To continue with anachronism, Kant was asking something more like, 'How is it possible to apply arithmetic in daily life, given that it is a pure science?' I suspect that Bertrand Russell may have been the first to see things in that way, partly because he had his own clear vision of 'pure mathematics' – the very two words with which he opens his first big book on mathematics (1903), as quoted in §2.11 above.

Kant's own notorious example was '5+7=12'. Aside from teaching and recitation in grade school, that proposition is most often used for what I shall call common-or-garden purposes: uses made by carpenters or shop-keepers, not to mention cooks making sure they have enough egg whites – five in this bowl, seven in that, as needed in this recipe for a large angel food cake. That is not what we now call pure mathematics.

I deliberately take an example in which one cannot tell, just by looking, that there are five egg whites in the bowl, after you have put them in. It is not like the bowl beside it, that contains five unbroken yolks, and you can tell there are five just by looking. And once you have separated out the whites, you cannot just count again. So you really do have to infer that in the two bowls of whites, taken together, there is the exact number of whites that the recipe calls for, twelve in all. This emphasizes the way in which 5+7=12 is used to make what Kant called a 'synthetic' judgement, which went beyond the judgements that one bowl contains five whites and the other seven (and that they are exclusive of each other, etc.).

26 Russell's version

When he was finished and done with *Principia*, Bertrand Russell sat down to write a pot-boiler that has charmed young people ever since: *The Problems of Philosophy*. It charms me still. It covers the philosophical waterfront, or at least that part about which Russell thought it 'possible to say something positive and constructive' (1912: v). In the middle of this little book he wrote that,

> The question which Kant put at the beginning of his philosophy, namely 'How is pure mathematics possible?' is an interesting and difficult one, to which every philosophy which is not purely sceptical must find some answer. (1912: 130f)

Russell exaggerated. Some great philosophies that are not purely sceptical have had no interest in mathematics whatsoever. But the question stands. What is so strange about mathematics that it should tie so many philosophers into knots?

Russell mentioned one source of wonder. 'This apparent power of anticipating facts about things of which we have no experience is certainly surprising' (1912: 132). Note the 'apparent'. Anticipating facts? If we know that if there are just five whites in one bowl and just seven in another, then we anticipate that there are twelve in the two bowls taken together and thus enough for the cake.

Russell's own example was not Kant's 'stock instance', 5+7=12, but 2+2=4. 'If we already know that two and two always make four, and we know that Brown and Jones are two, and so are Robinson and Smith, we can deduce that Brown and Jones and Robinson and Smith are four. This is new knowledge' (1912: 123). On the one hand, it is new knowledge, while on the other hand it seems to be acquired independently of experience. How, then, can a priori judgements anticipate experience?

Russell's example of Brown & Co. is uncharacteristically feeble. Far better to use F. P. Ramsey's illustration of the point in 1926 (Ramsey 1950: 2). From 'it's two miles to the station' and 'it's two miles from the station to the Gogs', 2+2=4 enables us to conclude that it is four miles to the Gogs via the station. (Ramsey was referring to his local topography – the Gog Magog Hills near Cambridge have a fine Roman road, a favourite for walkers, of which Ramsey was one. The hills are said to be, at 75 metres, the highest point of land across the great European plain until you reach the Urals.)

Russell argued that the problem of the a priori simply did not arise for pre-Kantian philosophers. Not, at any rate, if they thought, implied, or took for granted that such judgements are analytic, merely verbal, or some such. But being synthetic – somehow going beyond what is thought in the concepts of 'two' and addition – is not enough. It is the idea, which I do not find as explicit in Kant as in Russell, that the judgement can be used in everyday affairs.

In passing, note that Russell made no great play with the analytic/synthetic distinction. (As on many other topics, he changed his mind from time to time, including whether mathematics was synthetic or analytic.) In quite a number of instances, Quine's well-known theses explicitly repeat, with argument, what Russell took for granted. They were as one in their contempt for the idea of necessary truth and of analyticity. But Russell was careful to spell out the central idea of Kant's conundrum, that apparently we anticipate what we have not already experienced.

27 Russell dissolves the mystery

It was Russell who made Frege's analysis of number widely known (though only after reinventing the core idea himself), in his 1903 *Principles of Mathematics*. So we might expect him to give what we now think of as the Fregean dissolution of the puzzle about the apparent power of antici-pating facts. And indeed he probably does, allowing for the facts that (a) he was writing in what he hopes is an elementary and non-technical way, and (b) his vocabulary is no longer ours, and (c) we have developed a semi-technical vocabulary which we find expresses matters more clearly than anyone could then. And (d): Russell never stood still, which means in part that he was always changing in his mind.

Be that as it may, we get a neat series of inferences spanning some twenty small pages (the first edition of *Problems* really was a book you could put in your pocket). Russell embeds his discussion of a priori knowledge in a discourse of universals. He wryly remarks that, 'seeing that nearly all the words to be found in the dictionary stand for universals, it is strange that hardly anybody except students of philosophy ever realises that there are such entities as universals' (1912: 146). The argument proceeds briskly to the point that he can conclude: '*All* a priori *knowledge deals exclusively with the relations of universals*' (1912: 162, original emphasis).

'Thus the statement "two and two are four" deals exclusively with the universals concerned, and therefore may be known by anybody who is acquainted with the universals and can perceive the relation between them.' And 'it must be taken as a fact', that we can do this. 'We know *a priori* that two things and two other things together make four things, but we do *not* know *a priori* that if Brown and Jones are two, and Smith and Robinson are two, then Brown and Jones and Robinson and Smith are four' (1912: 164). To begin with, we must know there are such persons.

> Hence, although our general proposition is *a priori*, all its applications to actual particulars involve experience and therefore contain an empirical element. In this way what seemed mysterious in our *a priori* knowledge is seen to have been based upon an error. (1912: 165f)

There you have it. A philosophical problem neatly disposed of.

28 Frege: number a second-order concept

Frege insisted (against the 'formalists') that 'it is the applicability alone which elevates arithmetic from a game to the rank of science. So applicability necessarily belongs to it' (Frege 1952: 187). But as I read him, unlike Russell, he was not much worried by Kant's conundrum about the application of a priori knowledge. His analysis of number does provide an evident solution, and may be read as what Russell had in mind, albeit expressed in more careful logical language than mere hand-waving at 'universals'.

One of Frege's definitive achievements was to see that the statement, 'there are five egg whites in the bowl', should not be thought of as saying something about the egg whites, in the way that 'there are rotten egg whites in the bowl' does. It is best understood as, 'the number of egg whites in the bowl is five'. It is saying something about the concept 'egg whites in the bowl', and not saying anything about the whites, as we would be if we were saying they were rotten.

Hence the statement about the number of whites is what we now call a second-order statement, not about the egg whites, but about the concept of being an egg white in the bowl (a particular bowl, at a specific moment in time, being understood). To speak thus of *concepts* is, in scholastic parlance, to be 'intensional'; in Russell's equivalent 'extensional' way of speaking, it is to speak of the *class* of egg whites in the bowl, a class that has five members. Thus in extensional lingo the statement is once again second-order.

I think all this was understood by Russell, but it is common to say that a proper understanding of Frege came much later. Certainly we could not be sure it all hung together until George Boolos (1987) proved the consistency of Frege's second-order explication of arithmetic. The logical issues have been explored at length in Dummett (1991), Steiner (1998: ch. 1) and by many others.

You might think that Frege and Russell put paid to Kant's question. And so Frege, as elucidated by authors like those mentioned, did. Mathematics can be applied to common-or-garden situations because it is about concepts, not objects. But despite their adulation of Bertrand Russell, some members of the Vienna Circle tried to simplify or short-circuit the argument.

29   Kant's conundrum becomes a twentieth-century dilemma: (a) Vienna

There is a standard twentieth-century approach to the conundrum, deeply influenced by, but curiously tending to ignore, the Frege–Russell solution. We can think of it as a dilemma. The Vienna Circle took one horn and Quine the other.

Vienna took the two-sentence road. It taught that '5+7=12' is not 'a fact about items in the world', although there are many facts, often expressed briefly by the very same sentence. To change the example, consider the fact that the five loaves brought by Joseph to the picnic, plus the seven brought by Peter, make twelve in all. This fact is contingent (Peter is feckless and he might have eaten one on the way) and is checked empirically if need be. But

'5+7=12' is true in virtue of the meanings of the words, i.e. is analytic. Hempel (1945) is a classic exposition of the connections between the two propositions.

The youthful A. J. Ayer went to Vienna and came back a convert. In 1936, at the age of twenty-six, he used his wonderfully abrasive style to give the first side of the double-meaning idea its most colourful expression:

> Our knowledge that no observation can ever confute the proposition '7+5=12' depends simply on the fact that the symbolic expression '7+5' is synonymous with '12', just as our knowledge that every oculist is an eye-doctor depends on the fact that the symbol 'eye-doctor' is synonymous with 'oculist'. And the same explanation holds good for every other *a priori* truth. (Ayer 1946: 85)

That is an unwitting parody, but it was learned in Vienna. It is not a bad encapsulation of what, in careless moments, some of the Circle meant.

'Every other a priori truth': this is a primrose path to be taken with care. We know 5+7=12 by rote. This is possibly (certainly in a Montessori school, but in my own childhood not) backed up by the experience of putting five blocks together with seven and counting. The proposition that every prime is the sum of two squares, if and only if it is of the form $4n+1$, is very different, and not just because it begins with a universal quantifier (you can write the sum of 5 and 7 that way too).

Did Ayer really believe that the expression 'prime number of the form $4n+1$' is synonymous with 'prime number which is the sum of two squares'?

30 Kant's conundrum becomes a twentieth-century dilemma: (b) Quine

Quine demolished the concept of analyticity to his own satisfaction and that of a great many other philosophers. Hence he could not invoke two propositions, one synthetic and the other analytic, to explain the application of mathematics, either of the common-or-garden variety, or of theoretical physics.

There are not two propositions: there is only one sentence, '5+7=12'. So what is the one-sentence solution to the conundrum? Quine thought that mathematical statements can be called true or false only insofar as they have some real-world application. Totally 'pure' mathematics with no application can be called true only as a courtesy. He well knows that there is a

> vast proliferation of mathematics that there is no thought or prospect of applying. I see these domains as integral to our overall theory of reality only

on sufferance ... So it is left to us to try to assess these sentences also as true or false, if we care to. Many are settled by the same laws that settle applicable mathematics. For the rest, I would settle them as far as practicable by considerations of economy, on a par with the decisions we make in natural science when trying to frame empirical hypotheses worthy of experimental testing. (Quine 2008: 468)

The only mathematical utterances that are properly called true or false are ones that have application, and they are to be assessed as true or false only as part of our whole 'conceptual scheme'.

Kant's conundrum has been turned into a dilemma. The Vienna Circle grasps one horn: there are *two* propositions, only one of which is synthetic. Quine grasps the other horn: necessity is no more than pragmatic immunity. Only mathematics with application is true or false, and, if true, it is true in the same way that anything empirical is true.

There are many kinds of application, as we shall illustrate at tedious length in Chapter 5. What is the 'prospect of applying' Fermat's fact, that every prime of the form 4n+1 is the sum of two squares? Well, suppose I am set the exercise of arranging the blocks found in a box into two neat square arrays. For most boxes of blocks it is not worth trying. But I am shown that those in this box can be laid out in four groups each of n blocks, with one left over; I am told this is a prime number of blocks. Well, then I can start trying to make the two-squares arrangement with the blocks. If that is an application, let us call it Philosophically Pickwickian, after Mr Pickwick's kindly abuses of words to avoid offence.

Did Quine really believe that the proposition that any prime number of the form 4n+1 is the sum of two squares can be called true only 'if we care to'?

31   Ayer, Quine, and Kant

These are both great ideas. (a) There are two sentences. One is a priori, but merely verbal. The other is empirical. (b) There is only one sentence, but if it has no application, it is not even in the running for being true or false – except by decision, on sufferance. Each idea, in its day, captured a generation of analytic philosophers. Yet one may feel that both horns of the dilemma are too scholastic, too internal to a curiously linguistic form of analytic philosophy, to have satisfied Kant. Kant's problem, at least as glossed by Russell in 1912, is better resolved in the manner of Frege or Russell.

I have suggested that Russell made us see Kant's question, 'How is pure mathematics possible?', as not about pure mathematics but as about its application, in what I take to be the contemporary use of the idea of application. Kant did not put matters in that way because our distinction between the pure and the applied was somewhat different from his. I shall describe the emergence of these categories in Chapter 5A.

In my opinion, Frege and Russell resolved Kant's conundrum, as restated by Russell. But we cannot conclude that that's that. Questions about application continue. How come mathematics is so good at modelling the world?

32 Logicizing philosophy of mathematics

To borrow a bad pun from David Kaplan (1975), I have been Russelling Kant and even Plato. (Yes, that's 'rustle' as in stealing cattle.) This chapter has been, at bottom, a logicist answer to the question of why there is philosophy of mathematics *at all*. And, as already confessed in §3.4 above, I have redirected what Detlefsen said about how twentieth-century topics in foundations mostly arise from Kant. Mine is not logicism in the usual keys, but I am not able to escape my undergraduate training in moral sciences, based, to an extent that no one will credit today, on reading Frege, Russell, and Moore.

I call it a logicist answer, but it is also the answer from analytic philosophy, understood as a historical phenomenon. It matters not whether, as Michael Dummett (e.g. 1973) so forcefully contended, analytic philosophy has its roots in Frege, or whether, as so vigorously argued by Peter Hacker (1996), it stems from Moore and Russell. It may matter to what I am saying here that I happen to agree with Hacker. (My agreement is evidently due to upbringing, not argument.) My frequent allusion to Wittgenstein on mathematics is an allusion to someone whose mind was most intimately moulded by Russell, and who fought his duels with his internalized Russell in a logicist framework.

The logicist attitude to philosophy of mathematics – I venture to say the attitude of most analytic philosophers – is illustrated by the choices for anthologies. Benacerraf and Putnam (1964, 1983) dutifully insert a bit of Brouwer, Hilbert, and Bernays, but they composed two logicist collections. The same is in large part true of van Heijenoort's magisterial sampler, *From Frege to Gödel* (1965). One gains a very different picture from other anthologies, such as William Ewald's *From Kant to Hilbert* (1996) or from Mancosu (1998). One may then get the impression that *Principia Mathematica* was a sport (in the genetic sense of the word, an organism that differs markedly from other members of the same species, and might even give rise to a new species).

We have, then, two strands in the philosophy of mathematics, one 'logicist' and the other, for want of a better label, 'intuitionist'. (I do not mean the 'intuitionistic logic' axiomatized by Heyting, and whose semantics we owe to Kripke, but the attitude to mathematics epitomized by Brouwer himself.*) I beg fans of Hilbert to forgive me for this choice of labels. Hilbert and Brouwer are commonly thought of as opposed, but they shared much, and both had strongly Kantian roots and were deeply conscious of phenomenology. A clear distinction between Brouwer's 'intuitionist' and Hilbert's 'finitary' may have first been made by Paul Bernays (§7.3 below).

The two strands do not speak to each other very much. Certainly I do not speak to the strand of Kant that developed into 'intuitionism'. For a hint of how I think it goes, see Chapter 7A, a peripheral discussion of part of a single contribution by Paul Bernays. Happily Bernays is as good a protagonist of the 'intuitionist' strand as any – perhaps, in philosophical terms, better than any other. In §7.2 I quote the assertion that 'Paul Bernays is arguably the greatest philosopher of mathematics in the twentieth century.' I have no quarrel with that, but Bernays' topics are different from the ones that I discuss in this book.

If my question had been about the choice of problems arising in the twentieth century, I might well have developed the intuitionist strand and studied Bernays hard. But my question was about why there is philosophy of mathematics *at all*, why it is perennial, and I hope the answers given in this chapter may ring true, if incomplete.

33   A nifty one-sentence summary (Putnam redux)

'In my view, there are *two* supports for realism in the philosophy of mathematics: *mathematical experience* and *physical experience*' (Putnam 1975a: 73, original italics). I have avoided speaking of realism and will continue to do so, preferring to invoke the name of Plato when needed. Here I wish only to adapt Putnam's wise saying: there are two reasons why philosophy of mathematics is perennial. There is *mathematical experience*, and in particular the experience of proof. That is the Ancient reason, whose emblem is Plato. There is *physical experience*, the fact that we apply mathematics all the time. That is the Enlightenment reason, whose emblem is Kant.

34 John Stuart Mill on the need for a sound philosophy of mathematics

It is well known that John Stuart Mill held that the truths of mathematics are the most general truths of experience. Yes, in many respects, and not just this one, 'naturalized epistemology' begins with Mill. On a quite different topic, his notion of 'real Kinds' – predecessor of the notion of 'natural kinds' – was another instance of his naturalized epistemology. Whether or not a class constitutes a real Kind depends on how the world is. His approach is utterly opposed to the scholastic thought that sound classifications are grounded on essences.

He is better known nowadays for Frege's ruthless refutation of his empiricist conception of arithmetic than for what he actually said. Even less well known is his passionate conviction that it was essential to advocate his point of view, not only for philosophical but also for political and ideological reasons. Prose cannot get much stronger than this passage in his *Autobiography*, published in 1873:

> The notion that truths external to the mind may be known by intuition or consciousness, independently of observation or experience is, I am persuaded, in these times, the great intellectual support of false doctrines and bad institutions. By the aid of this theory, every inveterate belief and every intense feeling, of which the origin is not remembered, is enabled to dispense with the obligation of justifying itself by reason, and is erected into its own all-sufficient voucher and justification. There never was such an instrument devised for consecrating all deep-seated prejudices. And the chief strength of this false philosophy in morals, politics and religion, lies in the appeal which it is accustomed to make to the evidence of mathematics and of the cognate branches of physical science ... In attempting to clear up the real nature of the evidence of mathematical and physical truths, the 'System of Logic' met the intuitive philosophers on ground on which they had previously been deemed unassailable; and gave its own explanation, from experience and association, of that peculiar character of what are called necessary truths, which is adduced proof that their evidence must come from a deeper source than experience. (Mill 1965–83: I, 225)

Note that he was not objecting to the idea that intuitions – hunches – are important to mathematical exploration, conjecture, and research programmes. He was objecting to the idea that having an intuition was a sufficient ground for a claim to know something – or even a good reason for belief.

Mill saw the idea of necessary truth as a force for evil. He respected it greatly; he feared it; he hated it. His prime targets were the likes of William

Whewell. Whewell deserves a lot of credit for the reform of mathematics education in Cambridge, and perhaps even for the creation of the world's first Chair of Pure Mathematics (1863). But Mill saw an Anglo-Kantian who was carrying all the baggage of Kant's argot from the *Prolegomena*: 'a priori', 'apodictic certainty', and 'absolute necessity'. Worse even than that triad of monsters was something called 'intuition'! Between 1843 (*A System of Logic*), and 1873 (the *Autobiography*), Mill steadfastly regarded these ideas as pillars of reaction and of opposition to civil change.

Mill was happy to argue that aside from merely verbal truths, all knowledge is empirical and grounded in experience. The truths even of arithmetic are merely the most general inductions of all, confirmed in all our experience. I don't agree, but Mill's doctrine is serious philosophy, well updated, in a certain direction, by Philip Kitcher's naturalism (§17 above).

CHAPTER 4

Proofs

1   The contingency of the philosophy of mathematics

Chapter 3 diagnosed the perennial existence of the philosophy of mathematics. There are two strands in the answer, labelled Ancient and Enlightenment. One is located in proof, and more particularly in proofs of the more cartesian sort, that carry understanding and conviction with them. If proof is too strong an answer, it is the experience of exploration, which in our civilization ends in proof (among deeper sorts of understanding). The other is located in application. Here we argue that there could have been ample mathematics without human beings ever discovering demonstrative proof. We show in Chapter 5 that the distinction between pure and applied mathematics is recent, and the consequence of a pretty contingent sequence of events. Thus, to be crude, perennial philosophy of mathematics need not have come into existence. At most there would be specialist philosophy of mathematics, the way there is philosophy of physics or philosophy of economics.

In this chapter we recall Kant's famous story of how the very possibility of demonstrative proof was discovered in one place at one time by a very few people. It illustrates a theme common in human history. Discoveries of inherent capacities are made in one place and at one time by a handful of people in a small community. They can then spread across the face of the earth. Demonstrative proof – if one may be allowed to speak of it as one thing when in fact there is a motley of techniques that evolve over time – became and still is the gold standard, but it need not have been. The argument is that deep mathematics could have developed without what we call proofs at all.

Chapter 5 urges a less surprising claim about the common distinction between pure and applied mathematics. It is recent and might never have come into being. This is not a novel observation. Penelope Maddy (2008) gives her explanation of 'How applied mathematics became pure'. She holds that until the nineteenth century, mathematics was always applied mathematics, until an idea of pure mathematics hove into view. That is a viable description of the course of events, but mine is different. I hesitate to project the label 'applied mathematics' back beyond the period when it gradually came into use, namely after 1750. Moreover I focus more on verbal and institutional history than she does, and pick out different contingent events as central to what happened. I prefer to say that rather than applied maths becoming pure, our distinction between the pure and the applied came into being. It was prefigured by a distinction between pure and mixed mathematics, itself foreshadowed by an ancient distinction between pure and mixed science, but, not surprisingly, there were really significant changes in conceptions during this evolution.

Neither demonstrative proof nor the pure/applied distinction was inevitable. Both are contingent products of historical events, or, in scholastic jargon, simply 'contingent'. They could have been otherwise, or not come into being at all. Before proceeding we shall examine some lesser, more local, cases of contingency and inevitability. It will be useful to recall the Butterfly Model (inevitable development) and the Latin Model (a more random evolution) from Chapter 2, §§23–5.

A   LITTLE CONTINGENCIES

2   On inevitability and 'success'

It is obviously a merely historical fact that we have the branches of mathematics and the theorems that we do have. That's unexcitingly contingent, for we might never have found them out. Likewise for methods, concepts, and approaches. The questions might have interested no one. People might not have had the leisure to think about such things even if they wanted to. Or they might not have been smart enough to figure out what Fermat or Euler found out about primes of the form $(4n+1)$. Certainly the Langlands programme is something that arose in a particular context, and in a few decades it may be regarded as a byway (or on the other hand as a royal road).

Suppose that a community with interests very much like ours were on a mathematical walk, in something like the sense intended by Doron Zeilberger (§2.23), but a little more realistic, a little less 'random'. Would it inevitably get more or less to where we are now, if it succeeded in getting somewhere that was deemed to be rich and exciting and had many fruitful applications so that taxpayers would support it? The Butterfly Model of §2.23 says yes, the developments would have been inevitable.

Could this imagined community have achieved comparable 'success' some other way? The Latin Model of §2.25 says yes – all sorts of paths of development might have been taken, many ending up with satisfactory and interesting mathematics.

Unfortunately the very notion of 'success' is obscure. The maxim 'nothing succeeds like success' is barbed. Success is judged by what one has been able to do, and creates standards of what one wants to do. We might not have done so well by our present standards of success, but on a different walk we might have evolved other standards by whose measure our mathematics was altogether satisfactory.

I have tried to clarify the messy interrelations between 'success' and 'inevitability' elsewhere – but only for the case of more or less empirical inquiry, under the heading 'How inevitable are the results of successful science?' (Hacking 1999: ch. 3; 2000a). I do not want to rehearse those issues once again. I repeat only that the inevitability question leads to the question of what counts as success, which is itself strongly influenced by where we have got to, now.

I do believe that in the short run the developments of a science are pretty inevitable; here I differ from the more constructionist thinkers. I believe that in the longer run what we have done in a science was not at all inevitable, and in this I differ from the more scientistic thinkers.

I shall first mention two careful examinations of inevitability in mathematics. The first found that our notion of the infinite was not inevitable. The other judged that our notion of complex numbers was inevitable. The two propositions are formally compatible, although I suspect that many readers will welcome only one or the other, depending on their larger perspectives. I welcome both. This chapter is about neither. As announced, it will soon turn to the longer-term contingencies of proof.

3   Latin Model: infinity

Paolo Mancosu (2009a) asks, 'Was Cantor's theory of infinite number inevitable?' Cantor and his generation defined infinite cardinals – being equal in cardinal number for infinite sets – in terms of 1:1 correspondence. Mancosu traces a little-known tradition that extends another notion of equinumerosity into the infinite: two sets differ in number if one properly contains the other. A set is larger than any of its proper subsets. That's obvious, isn't it? Call it the subset model as opposed to the 1:1 model. Mancosu also cites Leibniz, who, on realizing that both options were open, opined that 'equal in number' is a concept that makes consistent sense only for finite sets, where the two concepts of equinumerosity coincide. Hence three walks were possible: Cantor's 1:1, the subset model, and Leibniz's cautious stroll, virtually standing still. (But Leibniz had a zillion ideas over the course of his life; we shall mention his idea of infinitesimals in a moment.)

If success is the development of an elegant intellectual structure and a rich research programme, then there is little reason to think that Mancosu's alternatives could ever be as successful as Cantor. But why should the imagined community not have different standards of achievement? What about having a lot of applications outside of mathematics itself? Many thinkers of very different persuasions say *that* is the measure of mathematical success. Cantor is not obviously successful in that way. In that light Cantor is *not* inevitable: we might have got on just fine without a serious transfinite, and used the proper subset model of equinumerosity.

But would we not have been intellectually the poorer? 'Less successful' in a larger scheme of things? Less successful by the standards of what in fact attracts the canonical great mathematicians? My purely personal opinion is that we would have been poorer, but I am prejudiced. I have loved Cantor's hierarchy of cardinals ever since I was exposed to it at a tender age.

If we begin in the mid-nineteenth century, with a great deal of mathematical analysis in place, and a desperate need for 'rigour' in the face of certain difficulties, then Cantor (1845–1918) was not out on a random walk. Especially once he had seen the richness of the terrain that he was opening up, there was one way to go. His insights and constructions were pretty inevitable, or so it seems to me. We can confirm this by the way that many figures that are still well known to us – Frege and Dedekind, for example – were also doing equinumerosity according to the 1:1 model. Some may judge that Dedekind (1831–1916) was far the more important contributor to the 'advance' of mathematical understanding than the philosophically more familiar Frege and Cantor. The very idea of a mapping – and, in some sense, the very idea of a function – is very plausibly attributed to him. Without such ideas, no Cantorian infinite. With such ideas, the Cantorian infinite may seem inevitable. Yet Kronecker (§§7.8–10) fought long and hard against that extension. Was it really inevitable that he would lose?

In the early days of the calculus, things were much less settled. Newton used infinitesimals, which long ago were replaced by the epsilon-delta foundation. William Thurston (§2.15) reminded us that in real-life teaching and learning the differential calculus, many understandings of the derivative are in play. Abraham Robinson showed how to make formally consistent sense of infinitesimals. Leibniz, he wrote, 'argued that the theory of infinitesimals implies the introduction of ideal numbers which might be infinitely small or infinitely large compared with the real numbers but which were *to possess the same properties as the latter*' (Robinson 1966: 2, original emphasis). 'It is shown in this book that Leibniz's ideas can be fully vindicated and that they lead to a novel and fruitful approach to classical Analysis and to many other branches of mathematics' (1966). Robinson used contemporary model theory to explicate this idea; subsequently a consistent formulation is provided by purely syntactic means (Nelson 1977). For one critique of the 'official' view that infinitesimals were doomed, see Błaszczyk, Katz, and Sherry (2013), and references given there.

If analysis had stuck to infinitesimals in the face of philosophical nay-sayers like Bishop Berkeley, analysis might have looked very different. Problems that were pressing late in the nineteenth century, and which moved Cantor and his colleagues, might have received a different emphasis, if any at all. This alternative mathematics might have seemed just as 'successful', just as 'rich', to its inventors as ours does to us. In that light, as Mancosu urged, transfinite set theory now looks much more like the result of one of Zeilberger's random walks than an inevitable mathematical development.

4   Butterfly Model: complex numbers

Meir Buzaglo (2002) addresses an issue similar to Mancosu's but reaches the opposite conclusion with different examples. He argues that most concept expansions are inevitable, in a rather profound way. His most familiar example is number itself, extended to include the field of complex numbers, not to mention the reals.

Many philosophers less iconoclastic than Wittgenstein have thought that the concept of number is strongly extended, modified, or changed when we speak of imaginary and complex numbers. Alice Ambrose (1959) thought Wittgenstein was in general wrong about the extent to which mathematical concepts are 'open-textured' (the useful phrase invented by Friedrich Waismann 1945), but that the successive extensions of the number concept were pretty free decisions, not determined by what was already in use.

Buzaglo argues to the contrary, and rather convincingly, that at the time that imaginary numbers were introduced as the solution to quadratic and cubic equations (nobody realized the scope of the concept expansion), this expansion was inevitable from within the concept of number itself. Many deep results in number theory (the theory of integers, or whole numbers) are proven using the complex field. It is as if number theory 'needs' complex numbers in order to prove facts that, in themselves, involve only whole numbers. An alternative but deeply unlikely hypothesis is that in the end we shall find 'elementary' proofs of everything, on the lines of Atle Selberg's achievement for the prime number theorem (§1.12).

Thus the expansion of the field of numbers to complex numbers is inevitable in the following sense. If mathematicians were to achieve the successes they have in fact achieved in proving results in number theory, they had to introduce complex numbers. Nobody knew this when the complex numbers were introduced, so their initial introduction was not inevitable. But in a slightly larger view, any theory of numbers would be impoverished if it did not become embedded in the complex field. To achieve our level of success with the theory of numbers, we have needed the complex field. This is a rather clear example of how the 'inevitability' question is entangled with the idea of 'success'.

The next step up from complex numbers is the invention of quaternions by William Rowan Hamilton (1805–65). It made classical mechanics far more tractable, and led on to an operator for the total energy of a system. Those operators, Hamiltonians, are standard for the physicists' toolkit. In a famous anecdote, the solution to a problem came to Hamilton in 1843, while on a long walk, and just as he was crossing a bridge. A bronze plaque on an insignificant Irish bridge marks the spot today. The incident, or the creation of the incident in memory, is utterly fortuitous, but on Buzaglo's type of account, the extension of the complex numbers to quaternions was inevitable. A contrary story has been related in impressive detail by Andrew Pickering (1995: chapter 4, 'Constructing quaternions').

This disagreement may be even more fraught. Hamilton thought that even the extension of positive integers to negative numbers needed justification. Number is quantity, after all, and negative quantity does not exist. Even a debt, negative in my accounts book, is a positive sum of money that I owe you. Hamilton (1837, written several years earlier) explained this extension in terms of time: negative number is just going backwards in time! He knew his Kant, and regarded 'algebra as the science of pure time', 'pure' in the sense of Kant himself. He put this together with his idea of couples, or what we now call ordered pairs. Hamilton invented that idea, which we now take for granted (Ferreirós 1999, 220f), and used it to introduce complex numbers as ordered pairs of real numbers. Complex numbers are thus governed by rules governing ordered pairs. Quaternions are ordered quadruples, and Hamilton found it a daunting task to work out rules for the algebra of quartets that preserved the central group-theoretic properties of the algebra of ordered pairs. The solution was the Eureka on the bridge.

Some readers, observing how he very carefully extended the concept of whole positive number to negative numbers, and then on to imaginary numbers, using ordered pairs of reals as the characterization of complex numbers, may feel that his invention of quaternions a few years later was a continuation of the same analogical reasoning. (More on Hamilton in §5.12 below.)

5 Changing the setting

Earlier work by Kenneth Manders (1989) and Mark Wilson (1992) suggests there is something wrongly posed in the 'debate' that I have constructed between Mancosu and Buzaglo. In both cases there was a change in the 'domain' or 'setting' within which mathematical concepts were entertained and played with. Wilson is concerned with (among many other topics in his rich paper) the way in which the complex numbers were reconceptualized with the development of projective geometry from the time of Jean-Victor Poncelet (1788–1867), who created the very framework in which Desargues' theorem (§1.9 above) could take its rightful place. Wilson holds that projective geometry provided a 'better setting' (his own scare quotes, p. 150: '"better settings" for old problems came into their own during the nineteenth century'.) Even more famous is Riemann's introduction of elliptic functions.

Wilson in effect defends what I call the Butterfly Model; a better setting could be compared to the metamorphosis from pupa to butterfly. 'The claim here is that the relevant mathematics achieves a simple and harmonious closure only when viewed in the proper – often *extended* – setting' (1992: 151). The proper setting – that implies there was really just one way to go.

Let us look more closely at the chain of events. Immediately after his graduation from the École Polytechnique in 1812, Poncelet was sent to the Russian front, as a lieutenant in a unit of engineers. Captured soon afterwards, he claimed to have written most of his *Traité des propriétés projectives des figures* (1822) while he was a Russian prisoner of war in 1812–14. This book quite literally opened up a new space, of which our intuitive Euclidean space is only a portion – we are like the inhabitants of Plato's cave, barely glimpsing the larger space. It is there that complex numbers find their home, and their relationships to the natural numbers become almost evident for the first time. They cease to be fictions that make no sense.

Let us not forget that Poncelet was an 'engineer': he will reappear in our discussion of 'pure' and 'applied' mathematics, §5.11. Ever a practical man, he was much impressed by the Russian abacus, or *schëty* (the Library of Congress transliteration of Счёты), which he brought back to Paris and encouraged his students to use (Schärlig 2001: 166). The first French equivalent to the British unit of power, namely 'horsepower', was the *poncelet*, about 1.3 hp. A man of wide interests and influence.

Wilson reminds us that popular histories teach the radical effect of Riemann's non-Euclidean geometry on our conception of space. But that was radical only with the hindsight of the representation of relativistic space-time that it made possible. Projective geometry enlarged space in a far more radical way. It also laid the basis for representing larger dimensions in two-dimensional drawings. Minkowski diagrams for space-time can be thought of as a continuation of the techniques of representation used in projective geometry. After that we have the conformal diagrams developed in the 1950s by Roger Penrose and Brandon Carter, often called Penrose diagrams. These enable us to present infinite four-dimensional relativistic space in two dimensions, and are essential tools of contemporary cosmological thought.

Manders and Wilson emphasize that the fundamental development in the birth of projective geometry was the creation of a new setting in which to think about spatial and algebraic relations. Once you have done that, what follows is pretty well determinate (the Butterfly Model). But the emergence of the new setting is not pre-determined (the Latin Model). Wilson even suggests that in the future new settings may emerge in which what was once inevitable will no longer seem right. I don't mean that truth goes to falsehood, just that what was an appropriate way of thinking at one time may pass from a certainty to a curiosity. (Some will want to impose the category of incommensurability here, but the cap does not quite fit.)

Extending space and stepping past the infinite may seem grand enough topics, but I wish to delve in a more fundamental way into what is contingent in the evolution of mathematics.

B PROOF

6 The discovery of proof

Proof has been a hallmark of Western mathematics descended from Greece. There was plenty of what we can recognize as mathematics in ancient Babylon, Egypt, China, and possibly in Mesoamerica. We might venture this: as soon as a people had invented writing, they were ready to invent mathematics. Just possibly that's the wrong order. Maybe people in Mesopotamia invented writing in order to keep track of numbers, as

Denise Schmandt-Besserat powerfully argues in *From Counting to Cuneiform* (1992). It is even possible that the Inca had quite sophisticated ways of keeping accounts without having developed writing, but using devices like *quipu* for records and the *yupana* for calculating.

Writing enabled us to tap cognitive skills in ways impossible without it. A handful of people in the eastern Mediterranean, Greeks as we call them, many from the coast of Asia Minor (modern Turkey), discovered the very possibility of deductive proof. They happened to live in an argumentative society, a few of whose members wanted a better tool for settling arguments than rhetoric.

I am trying to avoid the error of some great nineteenth-century historians of mathematics – an error that probably persists today – of suggesting that Greek mathematics was an 'immaculate conception' (to use an apt expression used in correspondence with Jens Høyrup, the distinguished historian of 'non-Greek' mathematics). Indeed I shall use the very excellence of ancient Chinese mathematics to imagine an alternative possible planet in which a Chinese, rather than Greek, tradition flourished, and was just as 'successful' a mathematics as ours, but with no emphasis on, or maybe no idea of, demonstrative proof.

Anyone with the least interest in reading older mathematics will see that standards of deductive proof changed over the years and are changing now. That is all contingent history. That does not need argument today. As I have emphasized from the start, at least in modern times we have had two ideals of proof on the drawing board, cartesian and leibnizian. In §3.11 we had a striking case of a proof written down by two humans, which no human other than they will ever check in its entirety. And that has (almost) nothing to do with computer-generated proofs.

I wish to argue the more striking thesis that the very idea of demonstrative proof seeming to convey apodictic certainty is, to use Zeilberger's natty word, something of a fluke.

7 Kant's tale

In the traditional story, relayed by Aristotle, a legendary Thales discovered proof. All the greats repeat what Aristotle passed on. My third epigraph, from Howard Stein, finds it 'significant that, in the semilegendary intellectual tradition of the Greeks, Thales is named both as the earliest of the philosophers and the first prover of geometric theorems'. I put Stein's words up front, not because of the legend, but because Stein sees the discovery of proof as the discovery, by humans at a particular time and place, of a human capacity. In his opinion, a similar discovery was made when the idea of a mapping was fully understood, only in the nineteenth century, largely through the insights of Dedekind.

A short time after Kant published the first edition of his *Critique of Pure Reason* in 1781, he caught the wave of the future and became something of a historicist about human reason. Yes: in the new introduction to the second edition of the first *Critique*, in 1787, Pure Reason has a History! But what, it may be asked, did Kant know about mathematics as practised in his own lifetime? Michael Friedman (1992) is a good place to go, and he emphasizes the extent of Kant's scientific acquaintance. One of Kant's major informants about current mathematics was the assiduous writer of new and up-to-date textbooks, Abraham Kästner (1719–1800), professor of mathematics and astronomy at Göttingen. Kästner was one of the first authors to write about applied mathematics, *angewandten Mathematik*, as described in §5.6.

I love Kant's 1787 rendering of the story of the light bulb going on over the head of Thales 'or some other'. Here I have blended the classic translation by Norman Kemp Smith (1929) with the recent translation by Paul Guyer and Allen Wood (1998). When the prose of the former is accurate enough and sounds better to me than the newer version, I have retained it. This is an occasion in which I have allowed personal taste to rule in blending two excellent translations. I have also broken Kant's very long paragraph into three.

> In the earliest times to which the history of human reason extends, *mathematics*, among that wonderful people, the Greeks, had already entered upon the sure path of science. Yet it must not be thought that it was as easy for mathematics as it was for logic – in which reason has to deal with itself alone – to light upon that royal road, or rather, to open it up. On the contrary, I believe that mathematics was left groping about for a long time (chiefly among the Egyptians), and that its transformation must have been due to a *revolution* brought about by the happy thought of a single man, in an attempt from which the road to be taken onward could no longer be missed, and by following which, secure progress throughout all time and in endless expansion is infallibly secured.
>
> The history of this revolution in a way of thinking – far more important than the discovery of the passage round the famous Cape of Good Hope – and of the lucky individual who brought it about, has not been preserved for us. But the legend handed down to us by Diogenes Laertius – who names the reputed inventor of the most insignificant steps in a geometrical demonstration, even those that, according to common judgement, stand in no need of proof, does at least show that the memory of the revolution, brought about

by the first glimpse of this new path, must have seemed to mathematicians of such outstanding importance as to cause it to survive the tide of oblivion.

A new light flashed upon the mind of the first man (be he Thales or some other) who demonstrated the properties of the isosceles triangle. For he found that what he had to do was not to inspect the figure; or even trace it out from its mere concept, and read off, as it were, its properties, but instead to bring out the properties that were necessarily implied in the concepts that he had formed *a priori*, and had thereby put into the figure he constructed. If he is to know anything with *a priori* certainty he must not ascribe to the figure anything save what necessarily follows from what he has himself put into it in accordance with his concept. (Kant, *Critique of Pure Reason*, Bx–xii)

Kant was of an epoch that favoured the Hero in History. He was committed to the excellence of the Greeks, 'that wonderful people', who were the fount of most wisdom. They owed far more to Babylonia and Egypt than Kant could know. And of course there was no first man, but rather some sort of community that became intoxicated with a new power, that of proving. Was it at the time of the legendary Thales (between about 625 and 545 BCE)? Most historians agree that the practices of making geometrical proofs crystallized later, perhaps at the time of Eudoxus (about 410–355 BCE).

The metaphor of crystallization suggests that a great deal of maths was around when it was all put together in a rather new way. The possibility of demonstrative proof was discovered, and after that Western mathematics was changed forever. For a wider use of the crystallization idea, as a phrase to capture historical transitions in thought, but falling short of the over-used 'revolution', see Hacking (2012a: §13).

Western philosophy was also moulded for a millennium after 'Thales', because demonstrative proof became the standard for real knowledge until finally Bacon, Galileo, and all the other usual suspects created another *revolution*. That is Kant's word. From the very beginning, the tradition was sure that something truly important had happened with 'Thales or some other'. 'Revolution', the word underlined by Kant himself, is used several times in a very few pages. As I. B. Cohen (1985) has emphasized, it was Kant, living in an age of political upheavals, who pioneered the concept of revolutions in science. Indeed he spoke of a first and a second scientific revolution, the first being mathematical and the second being what much later came to be called *the* scientific revolution. Kant interestingly identified that with the beginning of laboratory science, in which he emphasized using artificial apparatus to test well-thought-out hypotheses.

In both contexts Kant speaks not only of a revolution in thought and practice, but also of *discovery*. People found out how to do something. Our concern here is with the discovery of proof, not with the invention of laboratory science, but Kant had a very prescient view of both. Historians will protest that I am just repeating the usual claptrap. The usual claptrap may be weak history but it is important anthropology, a deeply serious claim about the discovery and exploitation of human capacities.

However we describe it, something tumultuous took place involving a very small number of individuals who corresponded and travelled in the eastern Mediterranean. Using the metaphor recently favoured in palae-ontology, Reviel Netz suggests 'that the early history of Greek mathematics was catastrophic'. 'A relatively large number of interesting results would have been discovered practically simultaneously' (Netz 1999: 273). He probably chose the word 'catastrophic' – in the sense of evolutionary biology, not catastrophe theory – because the word 'revolution' is almost worn out with over-use, but I think we can still see with Kant's fresh gaze.

8   The other legend: Pythagoras

'In the semilegendary tradition of the Greeks', Howard Stein recalled, 'Thales is named both as the earliest of the philosophers and the first prover of geometric theorems'. The only equal of Thales (?624–?546) in semi-legend is Pythagoras (?570–?495 BCE), a man who was about twenty-five when Thales died. Thales' home was Miletus in Asia Minor, then regarded as the richest of the Greek cities. Pythagoras was born on the island of Samos. Miletus was a short voyage away over a much-travelled inland sea. Today, Google Maps teaches, it is a 1,460 km trip by sea and land, since there are no ferries between the Greek island and the Turkish mainland.

Samos was once the technoscientific capital of the ancient world. The most memorable tyrant of Samos was Polycrates, who ruled 538–522 BCE, thus coming to power some eight years before Pythagoras departed for Sicily in about 530. He organized feats of engineering, such as the Eupalinian aqueduct on the island of Samos. This tunnel, over a kilometre long, and intended to carry water under a mountain, was built from both ends. The excavators from each side met tidily in the middle, with an 'error' of about 15 cm.

How do you build a tunnel under a mountain, starting from both sides and meeting in the middle? We would use triangulation. The engineers may not have known the Euclidean proof of what we call the Pythagorean theorem, but the facts about right triangles had been well known in ancient Egypt and Babylonia. It does seem that Eupalinos of Megara and his surveyors must have engaged in what we now call applied mathematics grounded in geometry.

Why was the tunnel built? Samos was in the heartland of ancient Ionia. The region bore the brunt of wars between the Persian Empire and the Greek city-states, with Egypt also being a major player in the region. So the cities of Ionia were constantly in actual or potential war, either internecine between the Greeks, or against empires to the east or south. Because of scheming and alliances, most often all three parties were involved in the ongoing battles, invasions, and general mayhem. The main port (now named Pythagoreion) had no ready access to fresh water: hence the value of an aqueduct from a spring on the other side of the mountain to enable the city to withstand a siege.

Herodotus thought the Samians remarkable in respect of engineering:

> they are responsible for three of the greatest building and engineering feats in the Greek world: the first is a tunnel nearly a mile long, eight feet wide and eight feet high, driven clean through the base of a hill nine hundred feet in height. The whole length carries a second cutting thirty feet deep and three feet broad, along which water from an abundant source is led through pipes. (Herodotus 1972: 228)

The other two Samian triumphs were the artificial harbour, whose break-water 'runs out into twenty fathoms of water and has a total length of over a quarter of a mile', and the 'biggest of all known Greek temples'.

Maybe you can build a great harbour without knowing much maths, but I cannot see how to dig a tunnel under a mountain, starting from both ends, without a rather high level of surveying skill. The water has to run downhill, so the exit must be a bit below the entrance, exit and entrance being separated by a kilometre and under a small mountain. All historians of ancient engineering will find this obvious, but philosophers who concentrate on the pure intellectual achievements of ancient Greece may need to be reminded.

In short, Pythagoras did not grow up in the provinces, but what was, in his early prime of life, the centre of a high-tech civilization. The island waned in importance, but much later both Epicurus and Aristarchus (who saw that the earth rotates around the sun) were born on the same island, and Aesop of the fables must be its best-known citizen.

9   Unlocking the secrets of the universe

We know precious little about the historical Pythagoras except his birth island. There is no reason to believe that he, the historical Pythagoras, proved, or even knew the proof of, the theorem named after him. Charles Kahn's *Pythagoras and Pythagoreans* (2001) is the best short compendium of what we know about his school. I made ample use of it in my own essay about both

Pythagorean and pythagorean mathematics, 'The lure of Pythagoras' (Hacking 2012b). To recall my convention, capital P Pythagoreans have a probable historical connection to the ideas of the Pythagorean school. Lower-case p pythagoreans are mathematicians and/or physicists with no connection to or descent from that school but whose metaphysics is in the same line of country: men such as P. A. M. Dirac, the great twentieth-century mathematical physicist who is a focus of that essay. With his 'Mathematical Universe Hypothesis' (§1.11) Max Tegmark affirms full-blooded pythagoreanism for the twenty-first century.

As Kahn writes (2001: 171), 'the last scientist who could be a genuine Pythagorean in the strict tradition of [Plato's] *Timaeus*, discovering the order of nature in the regular solids and the musical ratios', was Kepler. But legend still had it, well after Newton, that if only one had the texts of Pythagoras, one would be able to figure out the mathematical structure of the cosmos. A little more on modern pythagoreans in §5.17.

I propose that the very idea that we can reveal the secrets of the universe by doing mathematics originated with the Pythagoreans. Pythagoras personally grew up in a society that (I have just conjectured) had a good understanding of how to use calculation and surveying to achieve engineering feats. But the Pythagorean school were so taken with some mathematical phenomena – most notably, the harmonics of a string and the Platonic solids – that they took them to be insights into the nature of the universe, including solar astronomy, 'the music of the spheres', and the elements out of which the universe is built. These two interests prefigure the two concerns of physics as understood today, the very big and the very small, microstructure and the cosmos.

There is a certain irony about the role of Plato in this connection. For we know more about the original Pythagorean speculations from Plato's *Timaeus* than from any other text. Until the early modern period, that was virtually the only work of Plato's generally available, and every scholar knew Plato as the greatest of the Pythagoreans. Toying a bit with words, that is to say that he was once deemed to be the greatest exponent of 'applied' mathematics, rather than what he is today, the greatest propagandist for the 'purity' of mathematical reasoning and proof.

In Kant's book, Miletus is memorable as the home of Thales, and hence the site of the revolutionary discovery of the very possibility of mathematical demonstration. In my book, the nearby island of Samos was the site of high-level engineering mathematics, into which Pythagoras was born. And thence arose the Pythagorean dream of revealing the secrets of nature by mathematical investigations.

10 Plato, theoretical physicist

Walter Burkert was the leading Pythagorean scholar of the past half-century; Kahn (2001: 3) writes that Burkert 'has conclusively shown that the conception of Pythagorean philosophy that is taken for granted in later antiquity is essentially the work of Plato and his immediate disciples'. This is reciprocally true, for Plato's great work of Pythagorean cosmology is his *Timaeus*. Although Cicero translated it into Latin, the translation of a fourth-century Christian, Calcidius, was the one transmitted to mediaeval Europe, where until the twelfth century it was the virtually the *only* work of Plato's available. So the Plato who was known for a millennium *was* Pythagorean.

In the thirteenth and last book of the *Elements*, Euclid constructed the regular (convex) polyhedra, and proved that there are only five of them. Some scholars have (implausibly) argued that this conclusion was intended to be the pinnacle of the *Elements*, and was the motivation of the entire work. We call these five polyhedra the Platonic solids. We name them by the Greek words for the number of faces. Thus there is the pyramid or tetrahedron, the cube, the octahedron, the 12-hedron (dodecahedron) and the 20-hedron (icosahedron).

In ancient times these solids were arranged in order not of edges, but of vertices, namely four vertices (tetrahedron), six (octahedron), eight (cube), twelve (our 20-gon), and twenty (our 12-gon). Euclid provided constructions of each solid, in that order, and proved that the list was exhaustive. In *Timaeus* Plato presented the solids as the elements that constitute everything, namely, in order of the number of vertices, fire (composed of tetrahedrons), air (octahedrons), earth (cubes), water (made up of icosahedra). He said something obscurely cosmological about the last solid in this order, the dodecahedron or 12-hedron. Aristotle may have taken the aether to be built from dodecahedra.

There was plenty of pre-Socratic speculative physics, but just possibly Plato was doing something new. Today we might say (in the spirit of Eugene Wigner) that he took mathematics devised for supposedly 'aesthetic' reasons and applied it to speculative physics. But he did not think of things that way. He was a good Pythagorean: it might be best to say that he thought that in doing mathematics he was finding out about the deep nature of the world. In §6.13 we shall find Alain Connes saying that is exactly what we should do, now or as soon as we can.

The five Platonic solids are convex. There are non-convex regular solids: the star polyhedra whose history was constructed to great philosophical effect by Imre Lakatos (1976). Various generalizations of the basic structures are richly exemplified in nature. We have already encountered Michael Atiyah (§§1.12, 3.7). To get an idea of current interest in Plato's solids (generalized) it suffices here to quote the abstract of Atiyah's Leonardo da Vinci Lecture (2001) (Atiyah and Sutcliffe 2003). It gives these examples of a class of polyhedra that include the Platonic solids:

> The study of electrons on a sphere, cages of carbon atoms, central config-urations of gravitating point particles, rare gas microclusters, soliton models of nuclei, magnetic monopole scattering and geometrical problems concerning point particles. (2003: 33)

Atiyah also takes some delight in mentioning examples from the life sciences: 'spherical virus shells, such as HIV, which is built around a trivalent polyhedron with icosahedral symmetry' (2003: 34).

The Platonic solids have been fecund beyond belief, but have yielded their fair share of interesting but false inductions. The most famous is Kepler's modelling of the motions of the six known planets on the five regular solids. Kepler was a great Pythagorean, and thought he had explained why there are exactly six planets. This magnificent triumph of human ingenuity was based on a fortunate but mistaken induction.

Yet the persistence of Pythagorean phenomena is astonishing. I have just used Atiyah and Sutcliffe to illustrate the geometrical side. Then there was harmonics.

11   Harmonics works

Harmonics does turn out to be important to how the world works. It is not just music made numerological. But perhaps that is the wrong end of the stick. Because of the Pythagorean tradition of music and mathematics, students of nature deployed many harmonic series. Their role in physics may not be due to nature *qua* nature, but to the ways that we began to investigate nature on the Pythagorean model.

Every high-school student who learns any physics at all has to learn about simple harmonic motion, but is seldom if ever told why it is called 'harmonic'. Victorians got this better than moderns. William Thomson, Lord Kelvin, collaborated with P. G. Tait to produce the classic British textbook of Victorian physics, *A Treatise of Natural Philosophy*. Discussing simple harmonic motion, they wrote that 'Physically, the interest of such motions consists in the fact of their being approximately those of the simplest vibrations of sounding bodies, such as a tuning fork or a pianoforte wire;

whence their name; and of the various media in which waves of sound, light, heat, etc., are propagated' (Thomson and Tait, 1867: 1.i.§53).

Spherical harmonics were developed by Laplace, and are now used in characterizing the 'music of the spheres': for example, the magnetic field of planetary bodies, the cosmic microwave background radiation, and gravitational fields. Spherical harmonics are also integral to Dirac's papers relativizing quantum mechanics.

Is it just good luck that a tool developed from the ur-Pythagorean notion of harmony should have been so good at probing the universe? Or is it rather that we sense that the universe is 'harmonic' because of the tools that we use? Talk about 'the enigmatic matching of nature with mathematics and of mathematics by nature'!

Mark Steiner (1998) is the one contemporary philosopher to treat pythagorean matching of nature and maths with the astonishment that it deserves. He rightly uses the word 'anthropomorphic' in this connection, for it is a fact about human beings and their sense of structure (or whatever it is) that they project organization on to nature that nature surprisingly has. In §5.17 I shall briefly return to pythagorean thinking as one kind of applied mathematics. For the record, James Robert Brown, a philosopher of mathematics who is also a militant (but not strident) atheist, has written that 'Wigner's famous remark about the miracle of applied mathematics has been addressed (with religious overtones) by Steiner' (Brown 2008: 66).

12 Why there was uptake of demonstrative proof

Now let us leave Plato the Pythagorean and return to the more familiar Plato the purist. Netz makes plain that very few people were involved in ancient mathematics of the sort recorded in Euclid and to which Archimedes was the outstanding well-known contributor. (He has little to say about the Pythagoreans because he is interested in the explication of texts, not the piecing together of fables.*) Theoretical mathematics (to use Plato's own designation for which I do not much care) was a luxury enterprise. You had to have time and, among other things, money for 'postage', because your collaborators were often across the sea. Why then should there have been a cultural uptake of proof among Athenian intellectuals above any other place? Netz (1999: 299f) follows Geoffrey Lloyd (1990), putting it down to argumentative Greeks.

Athens was a democracy of citizens, all of whom were male and none of whom were slaves. It was a democracy for the few. But within those few, there was no ruler. Argument ruled. The trouble with arguments about how to administer the city and fight its battles is that no arguments are decisive. Or they are decisive only thanks to the skill of the orator, or the cupidity of the audience. Plato, fearful of democracy, capitalized on the mathematics of the few. For there was one kind of argument to which oratory seemed irrelevant. Any citizen, and indeed any slave who was encouraged to take the time, and to think under critical guidance, could follow an argument in geometry. He could come to see for himself, perhaps with a little instruction, that an argument was sound. Or so Plato urged on his fellow citizens.

By the way, we should record that our picture, of Athenians endlessly arguing and then getting impressed by proofs, reeks of the ivory tower, the glorious intellectual pilgrimage of the Greeks, inventing history, mathe-matical proof, and philosophy, as well as bequeathing the great tragedians to posterity. We who are not historians carelessly remember that the cities were constantly at war, but (we shrug) that is not their lasting importance. So it is well to recall that the chief reasons that the fractious Greeks were not overwhelmed by Persia were: (a) their naval power, and (b) that on land they had evolved arguably the best fighting machine of the age when used for combat in confined spaces. Persia had cavalry but the Greek cities had the hoplite unit, the phalanx. The hoplites were shield-bearers and the phalanx was a tightly knit unit in which each man was partially protected by the shield of his neighbour. Discipline, not argument, was intense.

13   Plato, kidnapper

Maybe the tale of some Greeks preferring proof to rhetoric is no more than a Just-So Story, but it is a lovely parable. Like so many parables it cuts two ways. In a brilliant critical notice of Netz's (1999), Bruno Latour (2008) speaks of philosophers after Plato 'kidnapping' mathematical proof, making it a cornerstone of his epistemology and metaphysics. A decade earlier Latour (1999) had already argued that Plato trumped Calicles' 'might is right', as a way to put down the demos, with a greater might, namely mathematics. That was in his essay on 'The invention of the science wars', which with characteristic eclat turns the tables. In order to illuminate those cultural epiphenomena of the 1990s, those 'wars', Latour wrote about 'Socrates and Calicles versus the people of Athens', and 'The triangular contest of Socrates, the sophists, and the demos' (Latour 1999: 219–35). (For a scholarly riposte, see Kochan 2006, but let us pass on to proof.)

> To the great surprise of those who believe in the Greek Miracle, the striking feature of Greek mathematics, according to Netz, is that it was completely peripheral to the culture, even to the highly literate one. Medicine, law, rhetoric, political sciences, ethics, history, yes; mathematics, no ... with one exception: the Plato-Aristotelian tradition. But what did this tradition (itself very small at the time) take from mathematicians? ... only one crucial feature: that there might exist one way to convince which is apodictic and not rhetoric or sophistic. The philosophy extracted from mathematicians was not a fully fledged practice. It was only a way to radically differentiate itself through the right manner of achieving persuasion. (Latour 2008: 445)

Latour speaks for myriads who say they hate or fear mathematics: '*If you have had to suffer through geometrical demonstrations at school (which is certainly the case for me)*' (2008: 443–4, my italics). I am in the minority. I went to a mediocre state school long ago, a rather unreformed place where I was lucky enough to be taught (slightly watered-down) Euclid at, in my case, the age of thirteen. Nerd that I was, I loved it: it was my escape from a harsher world around me. I learned about proofs, and delighted in them. Hence I am a gullible victim of Plato's abduction of mathematics, and also of Kant's Thalesian myth.

We must not, however, imagine that what dawned over the head of someone possibly called Thales was much like our notion of proof. That too had to be worked out. The punning title of Netz's book, *The Shaping of Deduction in Greek Mathematics*, alludes to the fact that diagrams play an essential role in ancient maths. One of Netz's triumphs has been to reconstruct the diagrams that are simply not there in the surviving texts. What evolved was a sort of formalism in the presentation of proofs, with many stock phrases (analogous to those found in Homer and in any oral tradition) and stock diagrams. That was a very specific conception of proof, different from ours, but it produced the sensibility that has been with us ever since: that some arguments are definitive, apodictic of themselves, and carry their own conviction with them.

14 Another suspect? Eleatic philosophy

Let us begin with a bit of history. While various types of computational methods have existed since antiquity – among the Egyptians, Assyrians and others – mathematics as we conceive it today was born in the fifth century B.C. in Greek ports, where people asked questions about philosophy or celestial mechanics and discussed such topics on the seashore. A certain number of these people, in an attempt to avoid self-deception, developed a discourse that would avoid confusion or misunderstanding, one whose very form would make it impossible to refute its content – a discourse that therefore would be convincing. In this context the notion of proof emerged in relation to the square root of 2. The Eleatic philosophers took this concept to a new level. Having established a discourse functioning locally, they wondered what was needed for this type of discourse to have greater validity, and this gave rise to the broader notion of proof. The idea of proof has taken various forms throughout history, but what they had in mind was absolute proof. (Connes, Lichnerowicz, and Schützenberger 2001: 23; henceforth CL&S)

Thus André Lichnerowicz (1915–98), a distinguished differential geometer and mathematical physicist, much influenced by the Bourbaki view of mathematics as the study of structure. We return to more of his ideas in §6.7 and thereafter. The Eleatics were a group identified with the colony of Elea in southern Italy, and usually included among the pre-Socratic philosophers. Notable among them are Parmenides, Zeno, and Melissus. We know them only from fragments and secondary sources. We certainly do not know enough to justify Lichnerowicz's 'history', but it is an excellent supplement (or antidote?) to the usual tale that features Thales and Plato. Note that in his story, the intention was to avoid self-deception, while in the account of uptake given by Netz and Lloyd the aim was to avoid rhetoric and deception by others.

Do not ignore as a mere storyteller's flourish that people 'discussed such topics on the seashore'. Anyone fortunate enough to have spent a year on a little-populated Mediterranean beach facing south will have noticed that the sun rises and sets on the horizon at a different point every single day, and that the movement is greatest between the solstices. Good for lesson 1 in celestial mechanics. It would never occur to anyone living where Paris or Calgary now stand.

We know least about the third Eleatic, Melissus of Samos. I have observed in §8 above that Samos, the birthplace of Pythagoras, was, for a few years, the high-tech capital of the ancient world. Melissus (who flourished in the fifth century BCE) seems to have held that the real is not *limited* in, or perhaps simply not *in*, space and time. That is, verbally at any rate, curiously similar to what twenty-first-century mathematical Platonists and platonists say about mathematical reality, mathematical objects, or abstract objects, depending on their choice of idiom. Lichnerowicz in effect regrets that the Eleatics are guilty of foisting upon Western civilization this notion of mathematical objects in their own realm of Being. We managed to escape that only in the nineteenth century when, thanks to the insights of Galois, we were able to focus on structures and not objects. Yet another myth of origin! Thanks to Galois, Lichnerowicz tells us, one of the obstacles to the development of Greek mathematics – the existence of mathematical 'objects' – was thereby overcome.

We thus realize that the nature of the entities under consideration is not important to the mathematician and that, in a certain sense, sets with isomorphic structure may be studied through a single treatment. Any set can in a certain sense be 'mathematized' if it obeys structural rules. Mathematicians have therefore learned that 'the being of objects' upon which they exercise their reasoning is of no importance. (CL&S 2001: 24)

Compare and contrast Latour and Lichnerowicz. Latour blames Plato for the fascination with apodictic certainty and proof, and thinks it is a *bad thing*. Lichnerowicz praises not Plato but his predecessors, the Eleatics, for discovering proof, which is a *good thing*. But he in effect blames them for their focus on mathematical objects, which he in turn thinks is a *bad thing*.

Of course none of this is history-as-fact; it was what we might call moral history or history-as-parable, which is often much more useful in philosophy than history-as-fact. (And, as a personal note to avoid misunderstanding, it is absolutely not 'archaeology' in the sense of Michel Foucault. His archaeology, which I have mimicked elsewhere, is among other things resolutely history-as-fact, but intended not to illuminate the past but rather to light up the present.)

15 Logic (and rhetoric)

I have quoted Kant with unusual admiration for his conception of the revolutionary discovery (by Thales or some other) of the human capacity for demonstrative proof. He did not find the development of logic interesting. 'It must not be thought that it was as easy for mathematics as it was for logic – in which reason has to deal with itself alone – to light upon that royal road, or rather, to open it up.' That's an odd phrase: in logic, reason has to deal with itself alone. His point is that in geometry one is dealing with something else, namely shapes and their relations, while logic reasoning has no particular subject matter: 'hence' it is less demanding.

On the central point, that mathematics has a subject matter – be it objects as was once thought, or structures, as many now believe – Kant was correct. Correct also to say that logic has no subject matter. And thereby hangs a Greek tale.

Socrates against the sophists, for example in *Gorgias*, insists that only subjects with a subject matter can be taught: carpentry, cookery, medicine, navigation. Since the sophists have no subject matter, they cannot be teachers.

There were, at the time, many teachers of rhetoric, and a great many manuals on the subject. Socrates was inveighing against such stuff. Hence it was an act of oedipal rebellion when Aristotle's first lecture courses were *Rhetoric* and *Dialectic*, the latter more often called *Topics*. Rhetoric, Aristotle tells us, is a counterpart to Dialectic: Rhetoric is for presenting an argument to an audience, while Dialectic is for arguing between two or more parties. 'It is possible to observe why some succeed by habit and others accidentally, and all would agree that such observation is the activity of an art (*tekne*)' (*Rhetoric* 1354^{a}9). He goes on to criticize the usual manuals of rhetoric, and states that he is not discussing oratory as such, but is providing a *theory* of good public speaking.

I emphasize that by general agreement among scholars, *Topics* and *Rhetoric* are the two oldest surviving works, doubtless courses of lectures. Only later did Aristotle invent the syllogism, as presented in *Prior Analytics* A 1–7. That is what I regard as the beginning of what we call logic, deductive logic, or of truth-reserving forms of argument.

16   Geometry and logic: esoteric and exoteric

I conjecture that the syllogism was immediately intelligible to the elite of the Greek city-states and abroad. This can be summed up as *exoteric*: the syllogism, exemplifying valid argument, was widely accessible to the elite. Nowadays we might say that Aristotle discovered the underlying idea of semantic logical consequence; less anachronistically, he discovered that some arguments are necessarily truth-preserving. He also set out the syntactic form of some such arguments, but does not appear to have emphasized this aspect of validity. At any rate, the syllogism became common coin: anyone in a society that uses writing can be got to understand the syllogism. Perhaps we should express this in reverse. In literate Greek society, understanding the syllogism became a minimum requirement for status.

This contrasts with the discovery described by Kant – the discovery of demonstrative proof in geometry. Geometry was practised by a small coterie of mathematicians scattered around the eastern Mediterranean, and had little influence on culture or society in general – except for Plato and his students. Then, as today, most persons of the highest status have no understanding of mathematical proof. Numeracy is required for social advance or even employment, then as today, although today the demand is now far stronger. But mathematicians of the calibre of Eudoxus and his pupils such as Theaetetus are a rare breed, and their pupils, who had even a minimum grasp and interest, were uncommon; their skills played no role in civic life. This is of course the thesis that Bruno Latour took from Reviel Netz, §13 above. I shall sum all this up in a word. *Esoteric*: geometry, with its demonstrative proofs, was for experts only.

Analytic philosophers may have had an incorrect idea of the historical relationship between mathematics and logic. They had almost nothing to do with each other until the middle of the nineteenth century. By the end of that century there arose the logicist thesis that mathematics *is* logic, pioneered above all by Frege. The culmination of his programme was the three-volume *Principia Mathematica* of 1910–12, by A. N. Whitehead and Bertrand Russell. This logicist thesis was of great interest, even though the results of Kurt Gödel are usually taken to show that it is false. This programme had an enormous influence not only on analytic philosophy but also on the teaching of logic in the English-speaking world. But connections between mathematics and logic simply do not exist until that epoch, and indeed most mathematicians paid little heed to logicism when it was being promoted. Thus the misconception, that logic and mathematics were always connected, is fostered by the accidents of pedagogy. In the English-speaking world especially, Aristotelian logic was replaced by what is called 'symbolic logic' patterned after the symbolism of *Principia Mathematica*. This symbolic logic was made exoteric, under the belief that it taught clear thinking, that it was doing what scholastic logic had always done, except better. Mathematics has, like the original geometric proofs, remained esoteric.

17   Civilization without proof

The discovery of the possibility of proof, and its uptake, is a sequence of almost accidental events: flukes of happenstance. We have, by good fortune, a historical illustration of that fact. China, in ancient times, developed brilliant mathematics, but it chiefly worked on a system of approximations. Proof seems seldom to have reared its head in China, and was seldom esteemed in its own right. Geoffrey Lloyd, who has devoted his mature years to comparative Chinese/European intellectual history, notes that the hierarchical structure of a powerful education system, with the Emperor's civil service as the ultimate court of appeal, had no need of proofs to settle anything. *That's* the critical contrast with argumentative Athens.

For a couple of hundred years – the whole romantic era and after – Europeans have extolled the 'glory that was Greece', to use the expression that Latour so ably makes fun of. German scholars led the way and did stupendous work. But there was an element of surrogate colonialism on the part of a Great Power that lacked colonies: high German civilization, up to

1933, presented itself as the continuation of the Greek one. Anything before that was barbarism.

The tide has happily turned, and scholarship on other civilizations is booming. For a vivid illustration, in the final months of 2010 New York University had a splendid exhibition entitled 'Before Pythagoras: The Culture of Old Babylonian Mathematics', organized by the historian of mathematics, Alexander Jones (see Rothstein 2010). Scholarship in ancient Chinese mathematics is likewise on the rise. There we have a terrific advantage over Babylonian studies, because we now have a generation of highly trained Chinese scholars to do most of the work. There have been two quite recent Sino-European editions of the most important text, commonly called the *Nine Chapters* (Shen *et al.*, 1999; Chemla and Guo 2004).

The title, 'Before Pythagoras', can be read two ways. The obvious intention is to say there was lots of mathematics before the notorious cult figure about whom we know precious little. But 'Pythagoras' connotes a fact, a theorem, and a proof. By the fact I mean the practical fact about the relations between the sides and the hypotenuse of a right triangle. By the theorem I mean the provable fact, and by the proof I mean, for example, the proof in Euclid.

One finds plenty of evidence of the practical *fact* (as opposed to the proof found in Euclid) summarized in Pythagoras' theorem in China, in Mesopotamia, and maybe even hints in Mesoamerica. Some preserved diagrams, which were on show at NYU, illustrate the Pythagorean fact. 'Before Pythagoras' does not mean before the fact, for the fact was probably known in many civilizations long before the discovery of proof. But it could mean 'before proof'.

There were well-understood facts, attested in experience, and accompanied by techniques for calculation and approximation: everything you could want, except demonstrative proof. And who needs that?

18 Class bias

Plato did not just kidnap mathematics; he abducted the whole of scholarship, so that everyone thought that theoretical mathematics *was* pretty much the whole of Greek mathematical practice, and (therefore?) the whole of ancient mathematics. Only now has it become generally recognized that there were 'two cultures of mathematics in ancient Greece' (Asper 2009). There was the esoteric culture represented to us by 'Euclid'. But there was also a vast amount of practical calculation. People did not go to Plato's Academy. They went to trade schools.

The same class bias has infected more than ancient scholarship. A great deal is known about the early Italian mathematical tradition with figures such as Tartaglia and Cardano and Fibonacci. With only slight anachronism this can be called a university tradition. Jens Høyrup (2005 and his website) has shown that there is also a trade tradition. In fact Fibonacci, whose name is immortalized in the Fibonacci numbers, was more from the trade than the university tradition. Italy was full of schools to which merchants sent their sons to learn efficient methods of calculation, Arabic numerals, and the Arabic *abbaco*, the (almost) hand-held calculator of the day. This was such a threat to the elite that in Florence, which we all know as the epicentre of the rebirth of civilization, Arabic numerals were forbidden and their use was penalized. And yes, scholars are recycling C. P. Snow's catchy phrase, 'two cultures' to speak of the two cultures of mathematics in the Italian renaissance, or what Høyrup (1990) calls the premodern practice of 'sub-scientific' mathematics.

Only recently have historians been able to undo their class bias in finding out who did what in the history of mathematical practice. (The same goes for many aspects of science.) Yet it is true that Plato kidnapped mathematics and made it into a practice of proof for many generations, including most 'pure' mathematicians alive today. 'Pure'? As we shall see in Chapter 5A, the very distinction between pure and applied mathematics is only a couple of centuries old – and Kant had a hand in inaugurating that distinction. The history of mathematics begins to look more and more like a history of events no more inevitable than the history of France.

19 Did the ideal of proof impede the growth of knowledge?

Plato's view on the importance of proof was turned by Aristotle into a generalized notion of demonstrative knowledge, to be called *scientia*. Popper taught that this was a terrible mistake, which held up the growth of scientific knowledge for a couple of millennia. Latour may seem a bit intemperate on the subject of Plato, but in 1945 Popper regarded Plato as the worst enemy of the Open Society, more dangerous even than Hegel and Marx. 'Socrates had only *one* worthy successor, his old friend Antisthenes, the last of the Great Generation. Plato, his most gifted disciple, was soon to prove the least faithful' (Popper 1962: 194).

Popper's idea was that what we now call scientific method was long impeded by the ideal of certain knowledge, demonstrated by proof. He is not obviously right. Maybe the scientific revolution, as we now call it, needed a prior model of demonstration in order to turn experimental exploration into laboratory science. In 1787, right after his discussion of 'Thales or some other', Kant rightly wrote of a new *discovery*, a second *revolution*, which in the twentieth century we came to call 'the scientific revolution'.

The 'royal road' of natural science could not be travelled along the road of demonstrative proof, and a new highway had to be built. Would we, as Popper averred, have done much better if we had never had that idea of proof at all? Or should we entertain a non-Popperian model, that without knowledge founded upon a conception of deductive proof, we would never have discovered laboratory science?

An unreconstructed thinker may well mutter: Popper's enthusiasms are all very well, but as a matter of fact laboratory science did not evolve in any tradition other than that of Greek proof. We owe everything to Plato's having set it up as the gold standard.

20   What gold standard?

I have been speaking as if there were just one thing, demonstrative proof, discovered by Thales or some other, and continued to the present day. I hope it is clear from Chapter 3, and the discussion of contingency thus far, that ideas and practices of proof evolve. Two successive studies of ancient Greek maths are my guides here. One is Wilbur Knorr (1975; also 1986). Knorr read the ancient texts so as to reveal new proof strategies; indeed he dated different bits even of Euclid in terms of a more or less linear development of proof ideas, which of course does not conform to the present order of the *Elements*. Oral history: the proof theorist Georg Kreisel, Knorr's colleague at Stanford, strongly encouraged this work. I believe he did so because it pointed not only to the history of proof but also to the human discovery of capacities for proving.

My second guide is Reviel Netz (1999), of whom I already made use in §4.8. He showed the role of diagrams and stock phrases as providing the logical superstructure of Euclidean proof. His primary enthusiasm is Archimedes, whose complete surviving texts he is editing. He shows how the Archimedean ways of settling things were a vast leap forward from the past. (He is prepared to mimic Whitehead on Plato, and say that all science is a footnote to Archimedes.) But then things slow down, and we have to move to other traditions, notably Persian and Arab, for new ways of finding out, in which the Greek ideal of proof does not loom large. Galileo may have mathematized the physical world, surely in an Archimedean way, but proofs are not so important for him. Arguably they become central again only after the new enthusiasm for rigour, often attributed to Cauchy (1821). I used Howard Stein in an epigraph, and requoted him above (§4.7) to go along with the conventional wisdom of Kant (and Einstein) about Thales or some other doing something radically new. But the very passage from which I quote begins with these words: 'Mathematics underwent, in the nine-teenth century, a transformation so profound that it is not too much to call it a second birth of the subject – its first birth having occurred among the ancient Greeks, say from the sixth through the fourth century B.C.' (Stein 1988: 238).

21 Proof demoted

I began this book asserting that the experience of proof is one of the two chief reasons that there is perennial philosophy of mathematics. Thus I began with the supposition that the idea of proof itself is a clear one. But very soon I was distinguishing cartesian from leibnizian proof: two different ideals of proof. One demands rather complete understanding after careful reflection, but is also connected with immediacy – Martin Gardner's 'Aha!' experience. The other demands systematic step-by-step checking. Only in the twentieth century did these two increasingly fall apart in ways that mattered.

In §1.26 I offered Voevodsky and Grothendieck as seeming extremes. Grothendieck demands ongoing conceptual analysis, not unlike the dialec-tic of Lakatos' proofs and refutations, which finally ends up with concepts whose relations are 'obvious' so that the theorem proved is immediately understood as true in a cartesian way. Voevodsky proposes not only more computational searching for proofs, but also that mathematical journals will accept theorems for publication only when accompanied by a machine-readable proof for electronic checking: the ultimate leibnizian situation.

In the course of this dialectic, set in motion early in the modern period – the time of what is called the scientific revolution in Europe – the simple idea of a demonstrative proof is 'demoted'. But there remains the experience of ideas falling into place, of concepts being clarified to the point where they demonstrably or even manifestly connect in necessary ways. This experience is deeply important to those who experience it, and creates the philosophical disputes that will be characterized in Chapter 6 by two protagonists, and more generally by a variety of options in Chapter 7.

22   A style of scientific reasoning

Here I may briefly interject a connection with another line of thought that I have pursued for some thirty years. The 'styles project', as I call it, is reviewed in Hacking (2012a). I introduced it as a study of styles of scientific reasoning in the European tradition. It turned a historical idea of A. C. Crombie's (1994) into a tool for philosophical analysis. It is important not to obsess on the word 'style', which has a lot of connotations, particularly the history of art. Mancosu (2009b) is an exemplary study of different styles of doing mathematics. I was using the word 'style' in a very different way, as a rather overarching concept for picking out the most general ways in which we think about a subject matter. I now speak indifferently of styles of thinking and doing, or genres of inquiry, or simply ways of finding out.

Following Crombie, I take it that a small number of fundamentally different genres of inquiry have emerged in the history of the sciences in Europe. The short names for six of these, as recognized by Crombie, are: mathematical, experimental exploration, hypothetical modelling, probabilistic, taxonomic, and historico-genetic reasoning. These are not sciences, for most sciences now use most of them; they are ways of finding out that have enabled human beings to dominate the planet, for better or worse. They are grounded in human capacities which have evolved in evolutionary time. These capacities were first exploited by a handful of people in one place and one time, but now have become part of the 'world system', to use a phrase employed by the anthropologist Marshal Sahlins. They emerged most distinctively in European history but have become a global phenomenon. They have become our standards of good reason.

The most unusual feature of the styles project is the claim that each genre of scientific investigation is 'self-authenticating'. Each style simply *is* our standard of good reason in a particular sphere of inquiry. It does not answer to some higher standard of truth. I shall not defend this perplexing doctrine here. But note that such an attitude is inhospitable to platonistic thinking. Proof does not answer to some body of truths not located in space or time or anywhere. It is our highest standard of truth in mathematics. I have argued above that we could have got on without proof as the gold standard, indeed without proof at all, but then mathematics would have been a very different kind of activity, far more akin to the empirical sciences than as it is practised today. It would not have employed the same style of reasoning. Not at all.

It will come as no surprise that in Chapter 6, which considers the views of two contemporary mathematicians, one avowedly Platonist and the other profoundly non-Platonist, it is the latter with whom, in general outline,

I agree. I remain deeply respectful of the other's position, in part because I find it an *interesting*, if unusual, variant on platonistic thinking. We shall see that neither mathematician is embroiled in present-day platonism, which is so deeply impregnated by semantics. The ideas considered by both are closer to Plato's than to those of most philosophers today.

This is not the occasion to develop the styles project, but I did want to say that my view of mathematical proof is part and parcel of this larger picture of human knowledge and discovery. Mathematics, in which proof is the gold standard of knowledge, is one of six distinct styles of thinking and doing that is an integral part of modern scientific activity. It is the one with the longest pedigree. Indeed just because what we now call pure mathematics makes no claims about how the world around us works, it is the most plausible candidate for being a self-authenticating method of inquiry.

CHAPTER 5

Applications

1 Past and present

This chapter continues the contingency theme of Chapter 4. Part A describes the emergence of the very distinction between pure and applied mathematics. The theme of the previous chapter – that we could very well have progressed without proof – can be debated. But there should be general agreement that we need not have divided mathematics into pure and applied. Indeed some decline to do so. Mark Steiner quite sensibly holds that there is mathematics, and there are its applications. Hannes Leitgeb, commenting on my first Descartes Lecture in Tilburg, said that applied mathematics is not mathematics at all, it is just science. As I understand him, the expression 'pure mathematics' is a tautology, for all mathematics is 'pure'. Steiner and Leitgeb have different motivations, but both reject the very concept of applied mathematics as standing alongside pure mathematics. I shall call this the Steiner–Leitgeb attitude to applied mathematics.

Here I take a different tack. I shall distinguish the pure from the applied, but then go on to say that they interact so vigorously that it often requires dogmatism to keep them apart. That will be illustrated by a few examples in Part B, 'A very wobbly distinction'.

A THE EMERGENCE OF A DISTINCTION

2 Plato on the difference between philosophical and practical mathematics

An important tradition in reading Plato on mathematics derives from Jacob Klein (1968). He argued that Plato made a fundamental distinction between the theory of numbers and calculating procedures. Here is a brief summary of the idea, due to one of Klein's students:

> Plato is important in the history of mathematics largely for his role as inspirer and director of others, and perhaps to him is due the sharp distinction in ancient Greece between arithmetic (in the sense of the theory of numbers) and logistic (the technique of computation). Plato regarded logistic as appropriate for the businessman and for the man of war, who 'must learn the art of numbers or he will not know how to array his troops'. The philosopher, on the other hand, must be an arithmetician 'because he has to arise out of the sea of change and lay hold of true being'. (Boyer 1991: 86)

We need not subscribe either to the terminology or the details of this interpretation to propose that ('real') mathematics, for Plato, did not include the arithmetic that we learned in school, and later applied in business transactions or ordering supplies for the troops. A redescription, owing more to Reviel Netz than to Jacob Klein, would be that Euclid's *Elements* made a decision to emphasize diagrammatic proofs rather than numerical examples. Hence, despite the depth of some work on numbers to be found in Apollonius and Archimedes, there was no tradition of 'advanced arithmetic' in antiquity in the way in which there was 'advanced geometry'.

Plato, then, put to one side the daily uses of arithmetic in technologically and commercially advanced societies such as Greece and Persia. Those uses are for what in the quotation are called 'logistic'. In my opinion we should avoid the notion that computation is for practical affairs in 'the sea of change'. That is the philosophical gloss of appearance and reality all over again. A primary point, closer to the experience of doing or using mathematics, is that computation is algorithmic. That is to say, it proceeds in a mechanical way by set rules. Calculations are not the sort of thing one needs to understand: one checks that one has not made a slip. There is no experience of proof as in the theory of numbers or geometry.

Probably there were plenty of manuals that taught how to calculate, complete with shortcuts, but as already suggested in §4.18, class bias has meant they are wholly forgotten and lost. They would be comparable to what we sophisticates dismiss as 'cookbooks', *mere* (!) 'how to do it' instructions that do not convey insight or understanding of 'how it works'. We may conjecture that that sort of text has not been preserved, partly because, on Platonic and then Aristotelian authority, it did not present 'science', *scientia*. Just possibly, if there had been a classical text of advanced arithmetic in antiquity, the questions of a priori knowledge, apodictic certainty, and necessity would not have been posed in that context. I like to imagine that those questions might, instead, have been *exposed* as pseudo-questions, at least in the context of the theory of numbers.

Plato (or his heirs) created a disciplinary boundary between mathematics, the science that every philosopher must master, and computation, the technique of commerce and the military. This bears some relationship to the recent distinction between pure and applied mathematics, but the fundamental difference is that the one involves proof, insight, and under-standing, while the other involves routine. It is the former that Plato 'kidnapped' while he abandoned practical mathematics by the roadside. Notice also that in Plato's vision (version Klein), there is far less of a motley of mathematics than in ours – because the routine computational side of mathematics is not (real) mathematics at all. Hence the Kantian exemplar, namely 5+7=12, would not have counted as mathematics for Plato (as understood by Klein and Boyer).

3   Pure and mixed

Francis Bacon was his usual prescient self when he devised that now abandoned term 'mixed mathematics', which appears at the end of the *OED* definition quoted at the start of §2.3. Gary Brown argues that this terminology is original with Bacon. His paper studies the evolution from a conception of mixed mathematics to a conception of applied mathematics current towards the end of the nineteenth century. As he summarizes it:

> More and more 'pure' mathematical theories of physics were developed to be 'applied' to more and more physically obtained data. By 1875 theories were no longer 'mixed' with experience, they were 'applied' to experience. (Brown 1991: 102)

Penelope Maddy has given another account of the evolution, especially in the nineteenth century, under the title 'How applied mathematics became pure' (2008). She particularly emphasizes events in Germany, and thus strongly supplements Brown, as well as providing a different and more philosophically germane type of analysis. What follows is another perspec-tive again; as is my wont, I am overly attentive to the words that are used, such as 'pure' and 'applied'. Also there is a little more use of historical incidents than is common in philosophy. The point is not to contribute to the history of mathematics but to illustrate the sheer contingency of the distinction between pure and applied mathematics. Had the history been different, our conceptual organization would have been different.

Bacon may be the origin of the term 'mixed mathematics', but it has ancient roots. The concept of mixed sciences is as old as Aristotle (Lennox 1986). It is integral to Galileo's way of thinking (Laird 1997). Yet what Yves

Gingras (2001: 407, n. 3) once wrote still appears to be true: 'There is a considerably large literature on the mixed sciences, but, to my knowledge, there is still no general account of its extension to fields other than the canonical ones (Astronomy, Optics, Harmonics).' Bacon made the old sciences central to what he was to rename 'mixed mathematics' (curiously omitting optics), but he did not stop there. He left the road open to a whole new category of knowledge.

Bacon was captured by the powerful image of the Tree of Knowledge. Every 'branch' (as we still say) of knowledge had to have its place on a tree. He needed a branch of the main limb of mathematics on which subjects such as mathematical astronomy could flourish, for they were mathematics in the older sense of the term, as noticed by the *OED*. These he called mixed, which also included perspective, harmonics, cosmography, architecture, and 'enginery'.

They were mixed because they 'have for their subject some axioms and parts of natural philosophy' (Bacon 1857–74: I, II). Mixed mathematics was not pure mathematics 'applied' to nature, but an investigation of the sphere in which the ideal and the mundane were intermingled. Note that in Bacon's tree, both the mixed and the pure were in that part of natural philosophy that fell under metaphysics, namely, the study of fixed and unchanging relations. But fairly early on there may also have been the thought that the relations examined in 'pure' mathematics could be studied just by thinking, and without a foray into experiment. Thus mathematics would be mixed if it demanded premises derived from experience. We have to be cautious, for there is a grave danger of falling into anachronism in these matters.

The Enlightenment was an era of classification, where natural history had a proud place as the science of nature observed. Hierarchical classifications, which we now conceive as branching trees, were the model for all knowledge, be it of minerals or diseases; the tree was also the model for the presentation of knowledge itself. Bacon's Tree of Knowledge was hardly new; that of Raymond Lull (1232–1315) was more graphic and may be thought of as the root of all Trees of Knowledge presented as diagrams. We now think of tree diagrams as one of the most efficient ways to represent certain kinds of information, but preserved tree diagrams begin to appear surprisingly late in human history (Hacking 2007). By Bacon's time they flourished in genealogy, logic, and many other fields. The Tree of Jesse – that is, the genealogical tree of Jesus Christ – was cast in wonderful glass on cathedral windows. That was the past: Bacon's own Tree of Knowledge was a benchmark for the future.

The image of a branching Tree of Knowledge was to persist for centuries after Lull and Bacon. It is most notably incarnate in D'Alembert's preface to *La grande encyclopédie*. It is a prominent pull-out page of Auguste Comte's *Cours*, the massive twelve-year production that continued the encyclopedic project of systematizing all knowledge and representing its growth in both historical and conceptual terms. The Tree of Knowledge, which was planted so firmly in the early modern world by Francis Bacon, has long been institutionalized in the structure of our universities with their departments and faculties. We no longer find pure and mixed mathematics on the tree, but many institutions have distinct departments or sub-departments for pure and applied mathematics – with mathematics often being filed as a branch of Arts and Sciences. (In older institutions it tends to have more status. It is a faculty unto itself at Cambridge. It is one of five basic fields of knowledge at the Collège de France, *mathématiques et sciences numériques*, which branches at present into five Chairs: Analysis and Geometry, Differential Equations and Dynamic Systems, Theory of Numbers, Equations in Partial Derivatives and Applications, and Algorithms, Machines, and Languages.)

D'Alembert's tree was vastly more structured than that of Bacon, largely because there were now so many branches of knowledge to accommodate. As conveniently diagrammed in Brown (1991: 89), there is now Pure, Mixed, and Physico-Mathematics. The subsequent structures are striking. Arithmetic plausibly divides into Number Theory and Algebra. One branch of Algebra is the infinitesimal dividing into Differential and Integral Calculus. But geometry divides into Elementary (Military Architecture and Tactics) and Transcendental (Theory of Curves). Military Geometry, e.g. ballistics, and the theory of fortification, seem to us to be perfect examples of applied, not pure, mathematics. The category of the mixed is, however, as we might expect, Bacon's list to which is added Optics, Acoustics, Pneumatics, and the Art of Conjecturing (see §5 below).

The arrangement of branches on the Tree of Knowledge is responsive both to the past and to innovation, but the basic arboreal structure has not changed much since the time of Francis Bacon.

4   Newton

What did Newton think about how to classify his work? He notoriously held his cards close to his chest. He presented what he knew, with few hints as to how he got there. My favourite example is that he got *G*, the gravitational constant, just about right, before he could possibly have had sound reasons. He is very close to the very good number that later workers deduced from Henry Cavendish's (1731–1810) ingenious experiment around 1796 to weigh the earth.

Newton's mathematics was, in a sense, wholly geometrical, whereas that of Leibniz and his successors was algebraic (another instance of the duality that puzzled me in Chapter 1). The very word 'geometry' was a common name in Britain for what we now call mathematics. Hence it is odd to read Newton quoted in Weyl (2009: 206): 'Geometry has its foundation in mechanical practice and is in effect nothing more than the particular part of the whole of mechanics which puts the art of measurement on precise and firm foundations.' There is no doubt that Newton's work, like much of Descartes', counted as mixed mathematics in the eyes of their contemporaries.

5 Probability – swinging from branch to branch

Many a new inquiry had to be forced on to the tree. Where, in the eighteenth century, would the new Doctrine of Chances, aka the Art of Conjecturing, fit? It was by definition not about the actual world, nor about an ideal world. It was about action and conjecture; it was the successor to a non-theory of luck. There was no branch on a Tree of Knowledge on which to hang it. Probability was uneasily declared a branch of mixed mathematics, less because of its content than because of its practitioners, such as the Bernoullis, who were mixed mathematicians par excellence. Consider, for example, that Daniel Bernoulli (1700–82) established the basic principles of 'hydraulics', which we now call fluid dynamics, and which deploys Bernoulli's theorem to this day.

The mixed, as we shall see, morphed into the applied. Hence the residual place for the 'theory of probabilities' as 'mixed or applied' mathematics alongside astronomy and physics in the *OED* entry. For some of the difficulties of sorting, see Daston (1988).

The Tree of Knowledge became the tree of disciplines. This may have a somewhat rational underlying structure, of the sort at which Bacon or D'Alembert aimed, but it is largely the product of a series of contingent decisions. This can be nicely illustrated by the location of probability theory in various sorts of institutions around the world. It was once a paradigm of the mixed, so you would expect it to continue as applied mathematics. That is certainly not what happened in Cambridge, where the Faculty of Mathematics is divided into two primary departments. One is Applied Mathematics and Theoretical Physics, the home of Newton's Lucasian Chair. The other is the Department of Pure Mathematics and Mathematical Statistics. Probability appears to have jumped from branch to branch of the Tree of Knowledge. In truth, to continue the arboreal meta-phor, it is an epiphyte. It can lodge and prosper anywhere in a Tree of Knowledge, but is not part of its organic structure at all. Maybe Bourbaki (§2.12) was right, after all, to exclude probability theory from its great, and rather tree-like, structure of mathematical structures.

6   *Rein* and *angewandt*

In §3.25 I said that the terminological distinction between pure and applied, *rein* and *angewandt* mathematics, was only just coming in when Kant was writing the first *Critique*. David Hyder drew my attention to the title of an influential book by Abraham Kästner: *Anfangsgründe der angewandten Mathematik* (1780). That is, *Elements of Applied Mathematics*. It is in two volumes. The first is an exposition of optics and mechanics – classic examples of mixed mathematics. The second deals with astronomy, geog-raphy, chronology (mechanical time-keeping), and *Gnomonik* (sundials). D'Alembert had filed these last two under 'geometric astronomy'. That all fell under mixed mathematics.

Kästner's book about what he actually called applied maths was the second instalment of a larger project, begun about 1760, and which finally ran to some eleven volumes, under the title *Anfangsgründe der Mathematik – Elements of Mathematics*. To judge by holdings in German, Swiss, and Austrian research libraries, the volumes on applied mathematics were the most widely used. Michael Friedman (1992: 75–8) shows that Kant knew Kästner personally and was familiar with some of his work.

Aside from his acquaintance with Kant, Kästner is a minor figure who is known for his lifelong search for a proof of the parallel postulate, and for his systematic exposition of fallacies in proofs that had already been proposed. Lobachvsky was a pupil of a pupil of Kästner's. Bolyai was taught by his father, who was one of Kästner's students. And Gauss, who matters so much in what follows, began his career at Göttingen when Kästner was still a professor there.

Kästner is surely responsible, if only indirectly, for the title of what is probably the first mathematics journal: the *Leipziger Magazin für reine und angewandte Mathematik*. Carl Friedrich Hindenberg (1741–1808) is a now forgotten probability theorist and, more generally, student of combinator-ics. Early on he was the tutor to a young man interested in mathematics. He accompanied his pupil to Göttingen, where he got to know Kästner (Haas 2008). Together with Johann III Bernoulli (1744–1807), Hindenberg pub-lished the Leipzig magazine for pure and applied maths for three years, 1786 to 1788, and an *Archiv der reinen und angewandten Mathematik* between 1795 and 1800. Many of the articles in the *Magazin* were about probability theory, including the theory of annuities.

In these two periodicals we have the first 'institutional' replacement of the pure and the mixed by the pure and applied, the *rein* and the *angewandt*. That begins in 1786, between the first (1781) and second (1787) editions of Kant's first *Critique*.

7 Pure Kant

Michael Friedman's *Kant and the Exact Sciences* (1992) is the classic contemporary exposition. Here we are concerned with a small matter peripheral to Friedman's interests and, for sure, to Kant's grand themes.

'Pure' – *rein* – evidently plays an immense role in Kant's first *Critique*, starting with its title. Quite aside from mathematics, the primary contrast for both the English and the German adjectives, then and now, is 'mixed'. In the *OED* we read 'pure' defined as follows: 'Without foreign or extraneous admixture: free from anything not properly pertaining to it; simple, homogeneous, unmixed, unalloyed'. For *rein* a few decades after Kant, Grimm's *Deutsches Wörterbuch* gives us: 'frei von fremdartigem, das entweder auf der Oberfläche haftet oder dem Stoffe beigemischt ist, die eigenart trübend'.

Hence Bacon's branching of mathematics into pure and mixed. The pure mathematics has no extraneous stuff added to reason; the mixed also has something empirical, or known only through experience. (The next, moralistic, sense of 'pure' – free from corruption or defilement, especially of a sexual sort – comes a close second: some have felt, over the ages, that mathematics is somehow corrupted by its applications, i.e. is no longer *pure*. One suspects that was the attitude of G. H. Hardy, §11 below.)

At the start of his rewritten Introduction for the second edition of the first *Critique*, Kant emphasizes what, for him, was the primary contrast: 'The Distinction between Pure and Empirical Knowledge' (B 1). The notion of 'applied mathematics' – or at least the phrase – was lurking in the wings. Thus in Kant's lecture notes, published as *Lectures on Metaphysics* (delivered from the 1760s to the 1790s): 'Philosophy, like mathematics as well, can be divided into two parts, namely into the *pure* and into the *applied*' (Kant 1997b: 307). This passage is from lectures probably given in 1790–91, four or five years after the initial 'institutionalization' of the distinction between pure and applied mathematics in the short-lived *Leipziger Magazin*. Kant, we may say, was *au courant* with contemporary changes in terminology, doubtless assisted by Abraham Kästner.

Passing from mere matters of words to matters of substance, Kant took arithmetic and geometry to be a priori conditions of all possible experience. This was a radical innovation, and yet a continuation of Plato's leitmotif. Kant was restoring a Platonic vision of mathematics as something utterly separate and absolutely fundamental to the nature of knowledge. He was also implicitly accepting Bacon's vision (by no means original with him, but so classified by him) of purity of knowledge as knowledge unmixed with, unsullied by, empirical content.

8   Pure Gauss

The separation of the pure from the applied, using those very words, was coeval with the first *Critique*, but not due to Kant. He did, however, set the stage for a German reorganization of ideas. The subsequent development of the very notion of pure maths, among German mathematicians, hinged on the ancient distinction between arithmetic and geometry. Purity was, to say the least, co-extensive with a priori knowledge and necessary truth. If any part of mathematics was deemed not to be a priori, but empirical, and/or deemed to be contingent, then that part could not be pure.

It is well known, to the extent of being philosophical folklore, that Gauss proposed that the Euclidean postulate be tested empirically by triangulation between three mountains. Geometry was synthetic, yes, but not synthetic a priori. In a letter of April 1817, Gauss wrote that, 'I come more and more to the conviction, that the necessity of our [Euclidean] geometry cannot be proven, at least not *by human* understanding nor *for* human understanding.' Perhaps in another life things will be different. 'Until then, we should not put geometry on the same rank with arithmetic, which stands purely *a priori*, but say with mechanics' (translated and quoted by Ferreirós 2007: 237; his square brackets). To put geometry with mechanics is to put it among mixed or applied mathematics.

Ferreirós observes, in the course of his beautiful analysis, that Gauss, in his inaugural lecture as director of the Göttingen Astronomical Observatory, expostulates against patrons and bureaucrats who demand that science be useful. They quite ignore the higher achievements of mankind – what makes us worthy as creatures on the face of the earth. The tone is quite similar to, though more literate than, those who today remonstrate about the way funding bodies demand that research have immediate application. Gauss spoke in 1808. Napoleon's armies still dominated the German-speaking peoples, and, as Gauss saw it, an interest in utility is characteristic of the evils of the day. Utility in mathematics was fostered by the loathsome French, to whom we turn in §10.

The German mathematical community, led by the incomparable Gauss, accepted the thesis that arithmetic is pure in the sense Kant intended, but geometry is not. This is a contrast between the nature of and grounding for two bodies of mathematics. Applied requires an element of experiential input, and so is not pure in the sense of Kant. This is different from our modern contrast, in which the pure is independent of practical purposes. Even if the two contrasts are relatively co-extensive, their meanings and motivations are very different. We have to go to France to witness what I take to be something more like today's concept of applied mathematics emerging.

9 The German nineteenth century, told in aphorisms

Here is one vision of the evolving mathematical world, from Gauss until the end of the century:

> Plato, according to Plutarch, said, *God eternally geometrizes.*
> Gauss said, in the first half of the nineteenth century: *God arithmetizes.* (Ferreirós 2007: 217f)

In his celebrated essay of 1888, 'What are the numbers and what should they be?', Dedekind (1996) used as his motto: *Man always arithmetizes* (Ferreirós 1999: 215f).

And then there was Kronecker's oft-quoted aphorism: *The good Lord made the integers, but all else is the work of man.* (Attributed by Heinrich Weber (1893: 15) to a talk Kronecker gave in 1886. For what Kronecker might have meant, see §7.10.)

10 Applied *polytechniciens*

The great French mathematicians of the generation that flourished in the era of the first *Critique*, men such as Lagrange (1736–1813), Laplace (1749–1827), and Legendre, did not see things as pure or applied. They were mathematicians. In general the scientists of that era made no such distinction. The distinction may have followed a different route in France from Germany. It involves some highly contingent events in disciplinary organization. The most important element may have been the French practice, going as far back as the late seventeenth century, of providing specialist engineering and artillery schools for the army. When the leading mathematicians of the day were hired to teach practical problems, our distinction between pure and applied began to take hold. So it is not surprising that in France we find the first fully institutionalized distinction between pure and applied. The distinction is not in terms of purity or the grounds of knowledge, but in terms of the point of what one is doing. That is how I understand the distinction today.

It is a useful rule of thumb in the history of science that when a discipline or set of disciplines becomes established, you should look for the foundation of specialist societies and journals. In 1810 Joseph Gergonne (1771–1859) founded the first long-running mathematics journal. Gergonne was Gauss's almost exact contemporary but a world apart, and not only in language and talent. His *Annales de mathématiques pures et appliquées* came out of the very world that Gauss feared.

In the prospectus for the journal, Gergonne begins by saying that it will primarily be devoted to 'pure mathematics', with special attention to what may simplify teaching (Gergonne 1810: ii). As far as he is concerned, geometry, including the new projective geometry, is pure mathematics, just like algebra and the theory of numbers.

The *Annales* will, however, also include 'applications to the divers branches of exact sciences', to wit, '*The art of conjecturing, political economy, military arts, general physics, optics, acoustics, astronomy, geography, chronology, chemistry, mineralogy, meteorology, civil architecture, fortification, nautical arts*, and *mechanical arts*'. That is pretty much d'Alembert's list of mixed mathematics. It also fits well with the way in which SIAM – the Society for Industrial and Applied Mathematics – defines its mission today, although a lot of new fields are added, including markets, finance, and the analysis of trading in derivatives (§16 below).

Gergonne, it should be said, founded the journal because he could not get a post at the École Polytechnique, where he had a feud with Jean-Victor Poncelet (§4.5; see §11 below) over projective geometry. Both, however, published articles on projective geometry in Gergonne's journal. Most of the articles in Gergonne's *Annales* were in fact contributions to projective geometry.

Although there had been the tentative Leipzig magazine for pure and applied mathematics, a permanent journal so named was not established in Germany until 1826, when A. L. Crelle (1780–1855) founded the *Journal für die reine und angewandte Mathematik*, the focus was rather different. Crelle published most of Niels Hendrik Abel's (1802–29) papers that transformed analysis. This is the oldest maths journal still running, although nowadays its articles are almost all in English.

As is the wont of French 'losers', Gergonne took a provincial post, in Montpellier. (Among well-known losers: Émile Durkheim in Bordeaux and Pierre Duhem in Lille.) A passage in J. S. Mill's *Autobiography* may dispel potential misunderstanding of what Gergonne was likely to be doing there: 'at Montpellier [writes the seventeen-year-old Mill] I attended the excellent winter courses of lectures at the Faculté des Sciences [... including] those ... of a very accomplished representative of the eighteenth century metaphysics, M. Gergonne, on logic, under the name of Philosophy of the Sciences' (Mill 1965–83: 1, 59. Mill's own notes on the lectures are to be found in 1965–83: XVI, 146ff). Gergonne comes out like a standard fin-de-siècle French *idéologue*. Mill calls that eighteenth-century metaphysics. Napoleon, who seems to have invented the label '*idéologue*' in order to trash those self-same late eighteenth-century metaphysicians, would have applauded.

For one of Gergonne's essays on the philosophy of mathematics published in his own journal, see Dahan-Dalmedico (1986).* Quine (1936: 103, n. 13; Benacerraf and Putnam 1964: 331, n. 13) states that the 'function of postulates as conventions seems to have been first recognized by Gergonne', citing an essay on the theory of definitions (Gergonne 1818), which he suggests as the origin of the phrase 'implicit definition', 'which has had some following', but which Quine rejects.

In fact 'philosophy of mathematics', so named, was a topic in Gergonne's *Annales* as early as the second volume. It was a hostile review of Wronski's *Introduction to the Philosophy of Mathematics* (1811), saying the book was hard to understand and used 'transcendental' vocabulary. I believe the book was reviewed in consequence of academic politics; Wronski (1776–1853) had sent some mathematics to the Institut de France, hoping to establish his fame. Lagrange, who was to referee, snubbed it; Wronski made a scene, and so, alas, on.

Wronski was a Pole who had settled in France, but only after serving in the Russian artillery (the book is dedicated to His Majesty the Emperor Alexander I, Autocrat of All the Russias). He travelled to Königsberg to study under Kant, only to discover that the great man had retired. I mention him only as a suggestion that his *Philosophy of Mathematics* may be the first book ever to bear this title.

Not that he is uninteresting to those who like eccentrics. He adopted a number of names – Hoëné de Wronski is the one that appears on his 1811 title page. He is said to be a source for Balzac's *La recherche de l'absolu* of 1834. The real-life anecdotes may be stranger than the fiction. In 1803 Wronski had an epiphany at a ball celebrating the birthday of Napoleon.

He understood that he, Wronski, would understand the Absolute via mathematics and thus lead man to redemption. When he was running out of inherited money, he made a deal with a banker that he would reveal to him the secret of the Absolute in exchange for generous support. When after sixteen years he did not deliver the Absolute, the banker sued for breach of contract, maybe 300,000 francs. (University teachers got 3,000 francs a year in those days.) Wronski appears to have won the case, convincing the judge he did have the secret but because of its nature could not reveal it. Crackpot, yes, but with ideas that enthusiasts or rag-pickers can later salvage to his credit, even if he was not smart enough to understand his own ideas: such as the belief that mathematics has to be founded on algorithms, and that the secret of the universe involves a relation between mass and energy. He did do a lot of competent but mediocre maths. There is a class of determinants named Wronskians.

11   Military history

In §4.5 we noticed Poncelet creating projective geometry while a prisoner of war in Russia. He was also the author of one of the first textbooks of 'industrial mechanics', published in 1829: the *Introduction à la mécanique industrielle, physique ou expérimentale*. That was a set of lecture notes prepared by one of his captains. The second edition (Poncelet 1839) is some 719 pages long. There was also a major print run for the third edition in 1870, not coincidentally the year of the war with Prussia. When I say Poncelet's captains, I mean it in the military sense. Poncelet rose to the rank of army general and commandant of the École Polytechnique, until he was fired by Napoleon III.

The inventor of projective geometry is an expert on industrial mechanics? A glance at the bureaucracy of war is needed. The École Polytechnique was created under the direction of the mathematician Gaspar Monge (1746–1814) in 1794. (In its first year of existence it was the École centrale des travaux publics.) Monge's forte was what is now called descriptive geometry: he devised plans for, among other things, fortifications. His system of projection solved problems of design so that instead of doing prolonged calculations, one could read off solutions from a diagram. Although this predates the projective geometry of §4.5 above, it is really the germ of the idea, and we can trace a (dotted) line from Monge to Penrose's conformal diagrams that enable us to represent an infinite four-dimensional relativistic universe on a flat piece of paper or a blackboard (cf. the end of §1.9). The first bit of the line is firm: Poncelet was Monge's student. And Monge, in his representation of fortifications, can be seen as redoing renaissance perspective. Indeed Morris Kline (1953) suggests that Desargues' theorem (§1.9) was prompted by an interest in perspective drawing.

To return to the institution, the present name, École Polytechnique, dates from 1795. Napoleon placed it directly under military control in 1804/5. Then and now it offered the best technical education in France and, for significant periods of time, in the world. It is still administered by the Ministry of Defence, although it was somewhat reorganized as a civilian organization in 1970.

In France an *école d'application* is an institution that provides specialist training for students who already have a 'post-secondary' education (as one says in English), attested by a suitable diploma, often a very demanding one. An *école d'application* is a school training graduates to apply their skills to a special field of expertise. The terminology is still in use for some of the *grandes écoles*, notably the École nationale d'administration, the breeding ground for the French politico-economic elite. But the terminology is very old.

A royal school of artillery was established at Metz in 1720, one of several across the nation. Metz is a historic garrison town in Lorraine, near the present border with Germany. These artillery schools, established at the end of the reign of Louis XIV, were regarded as the vanguard of science in the army, and the officers were required to have scientific/technical training. They were paralleled by schools of engineering, also with military functions. But the officers who went to these institutions were customarily members of the aristocracy. Thus Monge, who was merely the well-educated son of a merchant, and brimming with new ideas and enthusiasm, endured a long struggle with internal politics in order to win a position as instructor at the Royal School of Engineering at Mézières.

In 1794 the revolutionary Committee of Public Safety made the school at Metz an *école d'application* of what was to be renamed the École Polytechnique; that is to say, *polytechniciens* were sent there to become experts in gunnery. In 1802 Napoleon merged it with the nearby engineering college at Mézières, and it became the Metz École d'application d'artillerie et de genie (engineering), giving specialist training to graduates of the École Polytechnique.

Poncelet (1839) relates some of the antecedents of his textbook on 'industrial mechanics'. In 1817 he engaged to give classes to 'the artisans [*ouvriers*; this would include, e.g., architects and bridge contractors] of the town of Metz', and he published notes supplementary to his lectures. But then, charged, in 1825, with teaching, at the Metz *École d'application*, the course of Mechanics applied to machines, *I had to go more deeply into the theories that dominate this important branch of knowledge, and to make the study of their applications more easily accessible.* I thus became familiar with a way of looking at things which differs, in certain respects, from the ideas that are generally admitted in teaching elementary mechanics, and thus to approach more closely the methods adopted by the small number of geometers that have specially cultivated the science of machines. (Poncelet 1839; my translation, italics added)

This statement is relatively late in the day, but it illustrates the practical cleavage between pure and applied. Here is one of the greatest 'pure' geometers of the age. (French *Wikipedia*, accessed July 2012, called him exactly that!) He finds that he has to look at things differently when he is teaching young engineers in what is actually called an *école d'application*.

As a last reminder that this pure geometer had a job to do, take a look at his report on experiments made at Metz in 1834. They study the penetration of projectiles in various resistant materials, the rupture of bodies due to the shock of bombardment, etc. (Poncelet 1836).

12   William Rowan Hamilton

Hamilton, whom we have already encountered in §4.4, was one of the most fertile mathematicians of his era; some would call him unparalleled. I mention him here not for his genius but because he may have been one of the few mathematicians to speak, in exactly those words, of 'mixed or applied' mathematics as opposed to 'pure' mathematics. The use of just these words implies that at the time the English words were interchangeable; the terminology was thus in flux. That was in 1833. Brown (1991) reminds us that as late as the eighth edition of the *Encyclopaedia Britannica* (1853–60) there was an entry for mixed mathematics, which was changed to 'applied mathematics' for the ninth (1875–89).

Hamilton had been elected to the Chair of Astronomy at Trinity College Dublin at the age of twenty-two, while still an undergraduate; this was an ingenious way to provide the most brilliant man in the college (and probably in all Ireland) with a sinecure. That's only twenty years after Gauss, the far greater genius, had obtained what was (on paper, anyway) a similar job.

As an astronomer, Hamilton professed one of the oldest 'mixed sciences'. Unlike Gauss, he seems never to have spent a second on astronomy, but he did have a few obligations. His 'Introductory Lecture of Astronomy' is methodological, not a contribution to science, and apparently directed at the entering class at TCD, while serving as an inaugural lecture for his Chair. I have broken his florid paragraph in two, omitted a great many flourishes, and italicized the 'mixed or applied':

> in all the mathematical sciences we consider and compare relations. But the relations of the pure mathematics are relations between our own thoughts themselves; while the relations of *mixed or applied mathematical science* are relations between our thoughts and phenomena. To discover laws of nature . . . Induction must be exercised; probability must be weighed.
>
> In the sphere of the pure and inward reason, probability finds no place; and if induction ever enter, it is but tolerated as a mode of accelerating and assisting discovery, never rested in as the ground of belief, or testimony of that truth, which yet it may have helped to suggest. But in the physical sciences, we can conclude nothing, can know nothing without induction . . . Here, then, in the use and need of induction and probability, we have a great and cardinal distinction between the mixed and the pure mathematics. (Hamilton 1833: 76f)

The pure is a priori, and does not draw on experience. The applied is about phenomena, and how we may use mathematics to understand and change the world. This is pretty much the present understanding of the distinction between applied and pure mathematics.

This idea of the pure is a bit different from that which evolved among German mathematicians, convinced that geometry was synthetic and therefore not pure. Hamilton was a Kantian, but in the British mode. His idea is that algebra is the science of 'pure time' (Kant's *rein*), while geometry is the science of pure space. Interestingly he treated the continuum of real numbers as being grounded on the continuity of time, not space. 'The introduction to Hamilton's *Lectures* [1853, on quaternions] . . . presented quite interesting mathematical ideas under a dubious philosophical dressing' (Ferreirós 1999: 220). Dedekind read Hamilton, and incorporated among other things the idea that complex numbers are ordered pairs of real numbers (§7.9). But, writes Ferreirós, Hamilton's 'extreme reliance on philosophical ideas hindered the diffusion and acceptance of' his ideas (1999: 220). It might be better to say that it was Hamilton's reliance on Kant that (paradoxically) hindered his acceptance in Germany, where the action was to be. The likes of Dedekind, Kronecker, Cantor, and Frege were amply philosophical, but had withdrawn from the central doctrine of Kant's Transcendental Aesthetic, the duality of pure space and pure time.

13   Cambridge pure mathematics

Here I am not just insular but local. The analytic tradition in the philosophy of mathematics is properly traced back to Frege, but a lot of the stage-setting is grace of Whitehead and Russell. (Who would have heard of Frege, had not Russell told the world about his ideas?) Perhaps nobody really believed in Whitehead and Russell's great book, but even the great German set-theorists set themselves up with that work as a monument. Then it turned out to be essentially incomplete. So it is worth the time to consider the mathematical milieu that Whitehead and Russell took for granted – and their conception of pure mathematics for which they hoped to lay the foundations.

At least in British curricula we can locate the point at which 'pure mathematics' became a specific institutionalized discipline. In 1701 a Lady Sadleir had founded (in memory of her first husband William Croone, 1633–84) several college lectureships for the teaching of algebra at Cambridge University. In 1863, the endowment was transformed into the Sadleirian Chair of Pure Mathematics, whose first tenant was Arthur Cayley (1821–95). His tasks were 'to explain and teach the principles of Pure Mathematics and to apply himself to the advancement of that Science'. This was above all conceived to be algebra, and not geometric reasoning, in line with Lady Sadleir's original bequest, but now renamed.

Cambridge is undoubtedly more renowned for contributions to what we now call physics than to mathematics, insofar as they are distinct. The Smith's Prize, founded in 1768, was the way in which a young Cambridge mathematician could establish his genius. In the old days, up until 1885, it was awarded after a stiff examination in what would now be called applied mathematics. One wrangler (that is, one among the top in the examinations) who went on to tie for the Smith's Prize became the greatest British mathematician of the nineteenth century. But what we call mathematics has changed. We do not call him a mathematician but a physicist. I mean James Clerk Maxwell. Many names hallowed in the annals of physics, such as Stokes, Kelvin, Tait, Rayleigh, Larmor, J. J. Thomson, and Eddington won the Smith's Prize for mathematics.*

After 1863, what was called mathematics at Cambridge was increasingly pure mathematics rather than natural philosophy, which was reclassified as mathematical physics. It was within this conception of pure mathematics that Russell came of age. Likewise it was in this milieu that G. H. Hardy became the pre-eminent local mathematician, whose text, *A Course of Pure Mathematics* (1908, still in print), became a sort of official handbook of what mathematics is, or how it should be studied, taught, examined, and professed in England. (Wittgenstein wrote notes on the 1941 edition, which are being edited by Juliet Floyd.*) Russell's vision of mathematics was not determined by Hardy's, or vice versa, but the two visions are coeval, a product of a disciplinary accident in the conception of mathematics.

14 Hardy, Russell, and Whitehead

The master exposition of the glory of pure mathematics is G. H. Hardy's *A Mathematician's Apology* (1940). (Who knows the harm that this so elegant book has done to analytic philosophizing about mathematics?) Hardy was five years Russell's junior, and there are plenty of indications of Hardy's awareness, in his *Pure Mathematics*, of the foundational and conceptual worries that drove Whitehead and Russell in the first decade of the twentieth century. For example, in the first pages Hardy used a geometrical illustration to make us understand the Dedekind cut. In passing, he assured us that geometrical ideas 'are employed merely for the sake of clearness of exposition. This being so, it is not necessary that we should attempt any logical analysis of the ordinary notions of elementary geometry'. He wryly added, with a nod to Russell, that 'we may be content to suppose, however far it may be from the truth, that we know what they mean' (that is, have a clear conception of what the geometrical notions mean). Hardy may have written a book, but Whitehead turned the adulation of pure mathematics into a paragraph.

> The science of pure mathematics, in its modern developments, may claim to be the most original creation of the human spirit. Another claimant for this position is music. But we will put aside all rivals, and consider the ground on which such a claim can be made for mathematics. The originality of mathematics consists in the fact that in mathematical science connections between things are exhibited which, apart from the agency of human reason, are extremely unobvious. Thus the ideas, now in the minds of contemporary mathematicians, lie very remote from any notions which can be immediately derived by perception through the senses; unless indeed it be perception stimulated and guided by antecedent mathematical knowledge. This is the thesis which I proceed to exemplify. (Whitehead 1926: 29)

The glorification of pure mathematics by Hardy, Whitehead, and Russell has had, in my opinion, too much influence on analytic philosophy of mathematics. Those familiar with the local Cambridge scene will know that more important than the fact that Hardy, Whitehead, and Russell were all Trinity men, is the fact that all three were Apostles, a closed and indeed rather secretive Cambridge undergraduate club founded in 1820, and still going strong. So was G. E. Moore – and a number of memorable people at other times, including James Clerk Maxwell, whose talk to the club on 'Free Will' on 11 February 1873 is preserved as Maxwell (1882). And then there are the non-memorable 'Cambridge spies'.

There is a strong connection between the ethos of Hardy's *A Mathematician's Apology* and of *Principia Ethica*, which became the bible of the Bloomsbury group. It is obvious that Whitehead and Russell's magnum opus is named more after Moore's book than Newton's. It may be insufficiently noticed that the writing of *Principia Mathematica* exemplifies the same conception of moral obligation as does that of *Principia Ethica*.

15   Wittgenstein and von Mises

That world of Hardy, Moore, Russell, and Whitehead is the world into which Wittgenstein catapulted himself in 1911. What on earth did a man who had been studying aeronautical engineering, and building the prototype of a new kind of aeroplane propeller, make of the rarefied atmosphere he encountered when he knocked on Russell's door? He had been studying engineering in Manchester, where he lived in a rather drab flat. (W. G. Sebald's Max Ferber, an artist sent by his German–Jewish parents to England in 1939, rented a room in the house Wittgenstein had lived in thirty years earlier. Sebald includes a snapshot of it in *The Emigrants* (1996: 166f).) A drab flat in an industrial metropolis, but Wittgenstein had come from the esoteric atmosphere of the last great days of Viennese civilization: Klimt, the Mahlers (neither of whom he admired), and Musil, plus, a little later, Karl Kraus and Rilke (to both of whom he gave money). *Mutatis mutandis*, the Cambridge he encountered had the same pre-war self-confidence as Vienna, even if it had less talent. So perhaps he fitted right in to a little world of self-righteous self-confident eccentrics – insofar as Wittgenstein ever fitted.

Wittgenstein was a product of two civilizations, Cambridge and Vienna. So let us not attend only to the purity of the former before the war. There was a great debate about pure and applied mathematics even within the Vienna Circle.* The chief protagonist was Richard von Mises (not to be confused with his brother Ludwig, the economist). He is now best known to philosophers for his frequency theory of probability, a thorough work of, among other things, logical positivism. He himself strongly identified as an applied mathematician, and regularly insisted, against some other members of the Vienna Circle, that mathematics could be properly understood only by its applications.

Von Mises' dissertation was on the determination of flywheel masses in crank drives. I am sure it is just a coincidence, but Wittgenstein uses the simplest version of exactly this example a number of times (e.g. *RFM*: VII, §72 (a), 434). Von Mises was an aeronautical engineer, giving what I think was the first university lecture course ever, anywhere, on powered aircraft (Strasbourg, 1913). He became a test pilot during the Great War, but also designed the Mises-Flugzeug, a 600 HP flying machine, the most powerful and speedy craft ever, except it was too late in the war for development into a fighter plane (Siegmund-Schultze 2004).

Immediately after the war von Mises became head of the new Institute of Applied Mathematics in Berlin, and in 1921 he founded the *Zeitschrift für angewandte Mathematik und Mechanik*. Although most of the Vienna Circle were loyal to, or even formed by, 'Whitehead-and-Russell', and thus thought of mathematics in terms of pure mathematics, the residual effect of von Mises was strong, even though he himself had moved north. I imagine – but I have absolutely no documentary evidence for this – that this must have had an influence on a thinker of Wittgenstein's stripe. This might be especially so for a man who had been immersed in aeronautics.

In a fanciful mood I would suggest that it may be helpful to look at Wittgenstein stereoscopically, with one lens focused on Vienna, and the other on Cambridge. Through the Vienna eye one sees application. Through the Cambridge eye one sees purity.

16 SIAM

There are what I have already called 'common-or-garden' applications of mathematics. That is what Russell thought Kant's 5+7=12 was all about. Better to use Ramsey's example: from the fact that it is two miles to S, and two miles from S to G, we infer that it is four miles to G via S. But there are many more kinds of application than that. I shall presently propose a rather arbitrary classification of applications. First let us use the Society for Industrial and Applied Mathematics to exemplify one kind of applied maths, which could also, in a certain jargon, be called mission-oriented mathematics.

SIAM has 'over 13,000 individual members. Almost 500 academic, manufacturing, research and development, service and consulting organizations, government, and military organizations worldwide are institutional members.' It mostly does 'hard' or 'dry' applications. There is a lot more applied mathematics in the life sciences, and that field is growing incredibly fast, but not within SIAM. In a list of subdivisions of applied mathematics SIAM includes:

> 6. Applied mathematics and computational science is also useful in finance to design trading strategy, assist in asset allocation, and assess risk. Many large and successful hedge fund companies have successfully employed mathematics to do quantitative portfolio management and trading. (SIAM website)

As Hannes Leitgeb said after reading this out in his 2010 comments on the first Descartes Lecture, 'Nice job, guys.'

Since its founding in 1952, SIAM has increasingly focused on computational science. Here is its own current answer to the question, 'What is Applied Mathematics and Computational Science?'

> Applied mathematics is the branch of mathematics that is concerned with developing mathematical methods and applying them to science, engineering, industry, and society. It includes mathematical topics such as partial and ordinary differential equations, linear algebra, numerical analysis, operations research, discrete mathematics, optimization, control, and probability. Applied mathematics uses math-modeling techniques to solve real-world problems.
>
> Computational science is an emerging discipline focused on integrating applied mathematics, computer science, engineering, and the sciences to create a multidisciplinary field utilizing computational techniques and simulations to produce problem-solving techniques and methodologies. Computational science has become a third partner, together with theory and experimentation, in advancing scientific knowledge and practice. (SIAM website)

The first paragraph is pretty much a natural evolution from Gergonne's list of applied mathematics. The second is, as it states, 'emerging' with ever more powerful computation, some of which is still a twinkle in the eye of imaginative people (e.g. quantum computation).

B   A VERY WOBBLY DISTINCTION

17   Kinds of application

The institutional label 'applied mathematics' is usually clear enough in context. SIAM defines its mission with some precision, and so do most of the numerous journals with 'applied mathematics' in their titles. Likewise for university departments or sub-departments of applied mathematics, so named. But we may well ask, what gets called applied maths? Quite a few different things. Here are a few generic kinds of 'applications' of mathematics. The distinctions are not in the least sharp. I have numbered them, but the ordering is rather arbitrary, and does not pretend to be unique. I shall call the different kinds of applications by the silly name of *Apps*, simply in order to refer to them succinctly.

App 0: Mathematics applied to mathematics. That is where we started in Chapter 1, with, in the early modern period, Descartes' application of arithmetic to geometry, and continuing up into the present decade, with Ngô Bao Châo's proof of the Fundamental Lemma of the Langlands programme, which is an application of geometry to algebra. We observed that these 'applications' might better be described as the discovery of analogies between disparate fields, or of structural correspondences between them.

App 1: Pythagorean dreams. We broached these in the preceding chapter, §§4.8–11. There has long been, in a few minds, the sense that mathematics is uncovering the deep structure of the world – the idea that the essence of the universe just *is* mathematical. Plato was Pythagorean, and we know more about the aspirations of the Pythagorean cult from his *Timaeus* than from any other single text. The Book of Nature, as Galileo called it, is written in the language of mathematics. P. A. M. Dirac had a profound conception of this sort, expressed in many places, for example: 'There is thus a possibility that the ancient dream of philosophers to connect all Nature with the properties of whole numbers will some day be realized' (Dirac 1939: 129). Max Tegmark's Mathematical Universe Hypothesis (2008) is at present the most recent distinguished statement of the pythagorean philosophy (§1.11).

I do not share the pythagorean dream, but I respect it more than most. I think that the dream is absurd in itself, but that it may have a lot to do with the origin of the conviction that mathematics will reveal the Secrets of Nature. Many readers will be less charitable, and regard this vision of mathematics and the universe as lucky when it works, but puerile in inspiration.

App 2: Theoretical physics of the most general sort. Recent philosophical writing, in the analytic tradition, has chiefly discussed this kind of 'application'. I have in mind the kind of physics that invites a search for a Grand Unified Theory of everything. Physics dominated the philosophy of the sciences for the second half of the twentieth century, and typically it was high-energy physics, rather than condensed matter (formerly called solid state) physics. This is partly because of philosophical aspirations towards generality. Condensed matter physics tends to work with detailed models of specific phenomena, embedded within a general background of relativistic quantum mechanics; it has no ambition to tell the story of everything. Philip Anderson (b. 1923, physics Nobel Prize 1977) was the man who named condensed matter physics. In a classic paper, he expressed the philosophy of many a condensed matter physicist in three words: 'More is different' (Anderson 1972). I have never encountered anyone being pythagorean about condensed matter physics, whereas it is a strong temptation for some high-energy physicists. Yuval Ne'eman (1925–2006) is renowned for his 1961 co-discovery (with Murray Gell-Mann) of the 'eight-fold way' (SU (3) flavour symmetry). Ne'eman (2000) is a classic pythagorean take on twentieth-century mathematical physics. It is in this sort of context that Wigner's 'unreasonable effectiveness' is invoked. Chronologically, his type of astonishment at the success of mathematical physics is an affair that gets going early in the twentieth century, while Pythagorean dreams (*App 1*) are ancient.

App 3: *Mathematical modelling in 'disinterested' scientific research*. I am making a common but artificial distinction between research 'for its own sake' and research conducted to solve a relatively immediate 'practical' problem. Modelling has been radically transformed by the fast computer, so that now instead of laboratory experimentation one conducts computer simulations of numerous alternative models to see which best corresponds to known states of affairs. Many younger philosophers of the sciences now urge that laboratory experimentation is only one (material) kind of simulation, which is decreasingly needed in the sciences.

App 4: *Mission-oriented mathematics*. Most of what is actually called applied mathematics is deliberately done for a relatively practical purpose. Let us take SIAM as the prototype here. I believe that the expression 'mission-oriented' was a sort of buzz-word in US funding agencies, put into circulation in the 1970s, deliberately intended to favour practical work over 'mere' curiosity and speculation, which motivates *App 3*.

App 5: *Common or garden*. Then there are endless routine uses of mathematics by accountants, shopkeepers, carpenters, contractors, farmers, and lawyers. That is: by almost everybody in the contemporary world. Minimal numeracy is required for the minimum standard of living of an industrial state. It was doubtless developed for all sorts of practical purposes, although class bias has erased a lot of the memory of how that happened. Most of the elementary stuff, including Kant's 5+7=12, is best expressed not by propositions but by rules of calculation and procedure.

App 6: *Unintended uses*. All of the preceding are *intended uses* of mathematics, even when a branch of mathematics has been developed for one purpose and turns out to be useful for another for which it was not immediately intended. There are also *unintended uses*, where purists might say the use is almost a perversion of intended uses. I shall not return to these, so this entry will be more expansive.

For example, our educational systems force all children to acquire minimum mathematical skills, as demanded for a decent life today. But that is not the end of the matter. Those who do better at maths, at least up through high school, are advanced in the social hierarchy. As Chandler Davis* (1994: 137) has written, with a certain amount of venom, '20th-century mathematics surpasses Latin grammar, which was the sieve used to determine admission to the 19th-century British élite, and far surpasses the authentically applied mathematics which maintains its influence with other sciences and engineering.' One of the uses of such institutional practices is to preserve the present social order. The political scientist Andrew Hacker (2012) urges that this practice is disastrous for American education. Failing math is the main immediate cause of dropping out of US high school, he argues; with a different kind of requirement, far more young people would graduate with useful knowledge and skills. At the other end of the scale, elite medical schools require proficiency in calculus as a filter for their admissions, even though very few of their graduates ever make use of it. That is a beautiful example of Davis's thesis.

The most famous perversion of mathematics for elitist purposes was Plato's. Every senior civil servant in his Republic had to spend the best ten years of his life mastering mathematics: not in order to be a better finance minister or general, but in order to acquire a moral status. OK, maybe it is not a perversion. We could certainly do with politicians with more honour, and perhaps a ten-year mathematical apprenticeship would winnow out the dishonourable ones.

Consider also the mediaeval fascination with perfect numbers. (A perfect number is the sum of its divisors less than itself, as $6 = (3 + 2 + 1)$, or $28 = (14 + 7 + 4 + 2 + 1)$.) Euclid could find only four of them. Around the year 1000, Alhazen (born in Basra, modern Iraq) figured out a formula that generates many more even perfect numbers, but after the first few, the algorithmic computation grows *very* fast. As of 1 July 2012, forty-seven were known, and there is a massive computing project under way to find more. Are there odd perfect numbers? Are there infinitely many? We call all that pure mathematics. It seems to have no conceivable application. But some schoolmen were fascinated by perfect numbers because it might help them understand what perfection is. This is number theory applied to theology. Yet it is misleading to call it unintended, for that is exactly what the schoolmen intended to do. In a form of life a bit like Hermann Hesse's *Glass Bead Game* (*Das Glasperlenspiel*, also called *Magister Ludi*), first published in 1943, we imagine moral and aesthetic experience as the primary intended purpose of mathematical research.

That returns us to play (§§2.27–8), often dignified as 'recreational mathematics'. The element of play is certainly not unintended, although mathematical games often have unexpected consequences. Is it mathematics applied to recreation, like javelin throwing, once offensive, but now an Olympic sport? For one more game, see the end of §25 below.

App 7: Bizarre applications. Finally, we can imagine uses of mathematics that are not just unintended by the mathematicians who do the work, but, we might say, 'off the wall'. One of these is Wittgenstein's; it happens to be literally on the wall.

> Why should not the only application of the integral and differential calculus etc. be for patterns on wallpaper? Suppose they were invented just because people like a pattern of this kind. This would be a perfectly good application. (Wittgenstein 1979: 34)

It is possible that the wallpaper example was intended to soften his audience up for a much less bizarre application about which he had much more to say, namely predicting what a person (or most people) will do, when asked, say, to multiply 31 by 12. And that leads on to his reflections on following a rule, and much else that will not be discussed in this book.

Wallpapering the walls of a room sounds in the same league as tiling the floor. That takes us back into pure mathematics and even mathematical logic. The logician Hao Wang (1921–95) proposed that one way to attack the decision problem in mathematical logic was to rely on its analogy to problems of tiling a floor with identically shaped tiles of different colours, with restrictions on what colour can go beside what (Wang 1961). He showed that any Turing machine can be turned into a set of Wang tiles. All of the problems that he translated into tiling turn out to be undecidable. Tiling problems have become a growth industry in their own right. Pure or applied? Who cares.

18   Robust but not sharp

This classification into *Apps* is fairly robust, so long as nothing much is made to hang on it. As announced at the start of §17, the distinctions are certainly not sharp. Consider, for example, some of Hilbert's problems of 1900 (§2.21). The third problem for the 1900 Mathematics Congress (but no. 6 in the subsequent list of twenty-three problems) asked for a 'Mathematical treatment of the axioms of physics'. As Grattan-Guinness (2000) observed, it is not obvious whether to count that as imprecise, or as several problems. We can also ask whether it should be treated as a problem in pure or applied mathematics. I doubt that Hilbert would have bothered to answer. I think, but am not sure, that what Hilbert had in mind by problem no. 6 would best be filed under *App 2*.

One could suggest that a contemporary problem of this sort is to provide a rigorous account of renormalization in quantum field theory. (We saw in §2.21 that two physics-motivated problems occur among the seven Millennium problems, but not this one.) Renormalization has been essential in the toolbox of the theoretical physicist for decades. The practice was introduced in the 1940s to solve a difficulty that was manifest in the 1930s. A number of integrals at the heart of the theory diverge: that is, their value is infinity. (The theory in the first detailed instance was QED = quantum electrodynamics, though the difficulty appeared earlier.) So, to be crude and rude, we pretend their value is 1, and carry on. *It works* – to amazing experimental accuracy.

Physicists now take renormalization for granted. But it makes little mathematical sense. So it is a mathematical problem to produce an equally effective rigorous mathematics. Two authors whom we have already encountered – both Fields medallists – have taken up the renormalization challenge. One is Alain Connes (see Connes and Kreimer 2000; they have several related publications of the same vintage). A decade later another mathematician, Richard Borcherds (§3.12), essayed the same question in a more complex way (Borcherds 2011). One might think this is applied mathematics, in particular mathematical physics, *App 2*. In Borcherds' own opinion, 'It's mathematics rather than mathematical physics. It probably counts as pure rather than applied math. My own view is that it is pure mathematics strongly motivated by problems in physics' (email 14 December 2010).

As a curiosity, we observe that Connes and Kreimer (2000), 'Renormalization in quantum field theory', is published in *Communications in Mathematical Physics*, while Borcherds (2011), 'Renormalization and quantum field theory', appears in *Algebra and Number Theory*.

19 Philosophy and the *Apps*

Each *App* is connected to philosophy in its own ways. I find the inter-applicability of geometry and algebra astonishing, but others may not (*App 0*). It may diminish perplexity to say that there are very deep analogies between arithmetical and geometrical structures, but I still wonder, how come?

Mark Steiner (1998: esp. ch. 4) is a rare philosopher who takes the pythagorean *App 1* seriously. My curiosity is piqued also, but with less conviction that Steiner's. *App 2* is what prompted Eugene Wigner's musing about the unreasonable effectiveness of mathematics in the natural sciences. But he used stronger language than that in the course of his paper, in which he spoke of 'the miracle of the appropriateness of the language of mathematics for the formulation of the laws of physics'. Is it a miracle?

Steiner begins his book (1998: 13f) with a litany of similar cries of wonder from Heinrich Hertz, Steven Weinberg (*spooky*), Richard Feynman (*amazing*), Kepler, and Roger Penrose. I myself would not go further than 'amazing'. Penelope Maddy (2007) argues that the phenomenon is not even amazing. We should regard it as straightforward modelling in which there is a lot of groping and grasping and trying out of mathematical models. From this point of view, *App 2* is no more than a rarefied case of *App 3*, rather than a contemporary continuation of *App 1*.

That is a very sensible approach for anyone who dislikes spookiness. Nevertheless it may not address, to everyone's satisfaction, the curious luck that nature has bestowed on some pythagorean speculations. Something merely attractive to the human mind – certain symmetries, to use a recent example – turn out to unlock some of the secrets of nature. A human sense of 'fittingness' sometimes pans out better at analysing the universe than it ought.

In the case of *App 4*, SIAM surely got it right as quoted above: 'Applied mathematics uses math-modeling techniques to solve real-world problems.' That is: what we need is an understanding of models, not in the sense of semantics, but in the sense of the physics, first brought to the fore in philosophy of science by Cartwright (1983), and now an active research field in its own right. That is a widely held attitude – application is all models. It is not quite as simple as it sounds. Wilson (2000) mocks Wigner's title with one of his own, 'The unreasonable uncooperativeness of mathematics in the natural sciences'. I shall pursue modelling further in §21.

I claim that *App 5* prompted Kant's and Russell's problems as discussed in Chapter 3B, and, with less confidence, claim that the Frege solution works rather well. I don't suppose that *App 6* brings up any special philosophical issues. It is just a reminder of unexpected uses and abuses of mathematics. Almost anything, including mathematics, can be put to many unintended uses, however useful or silly they may be. Finally, I mention *App 7* only because Wittgenstein makes such astonishing use of weird examples.

None of these philosophical issues connected with the application of mathematics would have arisen in anything like their present form without the distinction between pure and applied. Thus what I have called the Enlightenment strands in the philosophy of mathematics also depend upon a highly contingent history.

20   Symmetry

Just a few words more on the pythagorean dream. Yuval Ne'eman, whose greatest discoveries involved symmetries, takes symmetry to be a profoundly pythagorean idea. And so it is, for the Platonic solids exemplify many types of symmetry, which may well be why polyhedra are so useful in modelling so many different natural phenomena. Mark Steiner rests a good deal of his pythagorean case on the power of symmetry in late twentieth-century fundamental physics.

Marcus Giaquinto, in commenting on my third Descartes Lecture, took the human recognition of symmetry to be one of the bases, roots, or origins of mathematical thinking. Deep in human nature is a sense of symmetry, evident in so many early artefacts, and intensely investigated by cognitive psychologists today.

We could combine the thoughts of Giaquinto and Steiner to make this suggestion: humans have a deep sense of symmetry. Symmetry is fundamental to many natural processes. One of the causes of 'the enigmatic matching of nature with mathematics and of mathematics by nature' (§1.1) is that this 'anthropological phenomenon', mathematics, derives much from the innate human appreciation of symmetry, and the fact that Nature herself is full of symmetries. This in turn would be grist to Penelope's Maddy's naturalist mill: there is nothing spooky about this, and it provides no basis for a pythagorean philosophy of mathematics.

There are, however, two anomalies of very different sorts. One was alluded to in §2.22. Hon and Goldstein (2008) argue that the modern concept of mathematical symmetry did not exist before Adrien-Marie Legendre, in his 1794 reformulation of Kant's use, in 1768, of incongruous counterparts to prove that space is absolute (Kant 1992). Hon and Goldstein argue that the recognition of symmetry by Legendre is a watershed in the history of mathematics and of all the sciences.

Yes, they say, there are ideas of symmetry in Euclid, and also there is an aesthetic concept of symmetry, referring to pleasing proportions. Euclid used the word *summetria* when two magnitudes share a common unit of measure. Thus the diagonal of the unit square is incommensurable, or 'asymmetric', with its side. Neither the proportionality nor the commensurability idea is a precursor of Legendre and all that followed. That includes the symmetry of which Yuval Ne'eman and Murray Gell-Mann (and many others) made such good use in creating the Standard Model, now so entrenched in contemporary physics. I am inclined to argue against Hon and Goldstein. The co-option of the word 'symmetry' by Legendre in 1794 does not, in my opinion, vitiate the claim that there is an ur-idea of (what we call) symmetry running from ancient times to the present. In the light of Hon and Goldstein's impressive research, however, this needs to be established.

Another fact mentioned in §2.22 is also curious: Joe Rosen's observation (1998) that his own primer (Rosen 1975) filled a decades-long void in expositions of the mathematics of symmetry, whereas today there are almost uncountably many. That is a mere curiosity (I think). Of greater interest is the question of symmetry and recent fundamental physics.

The sense that symmetry was the inevitable way things had to develop in post-1950 physics is very powerful. But often inevitability is a misperception. Recollecting the same period of development, Steven Weinberg (b. 1933, physics Nobel Prize 1979) has written: 'As I recall the atmosphere of particle physicists towards symmetry principles in the 1950s and 1960s, symmetries were regarded as important largely for want of anything else to think about' (Weinberg 1997: 37).*

21   The representational–deductive picture

Philosophers of very different stripes offer what is almost a consensus view of applied mathematics, especially of type *App 3* but also of types 2 (theoretical physics) and 4 (SIAM). I shall call it representational–deductive. This label is patterned after familiar talk of the hypothetical–deductive method in the natural sciences. (That phrase itself stems from William Whewell (1794–1866), and his great work of 1840, *The Philosophy of the Inductive Sciences, founded upon their history.*)

Here is the picture. We are interested in some phenomenon. We try to form a simple abstract model of the phenomenon. We represent it by some mathematical formulae. Then we do mathematics, deductive pure mathematics, on the formulae, in order to try to answer, in simplistic terms, some practical questions about the phenomenon, or to understand how it works. Then we 'de-represent': that is, translate a mathematical conclusion back into the material phenomenon.

The most important mathematical or physical event in early modern science was the understanding of acceleration: the idea of rate of change of

change. That was a truly incredible breakthrough. Change is easy to understand: rate of change was incredibly difficult to make sense of. The achievement of mastering it prompts the phrase 'mind-boggling': the human mind
made a great leap forward.

The differential calculus was the immediate result. To this day, partial
differential equations are at the core of much (most?) applied mathematics.
There is no 'unreasonable effectiveness' here. Things move. Seventeenth-
century natural philosophy was about motion. Rates of movement change.
I am quite unfashionably glad to speak of heroes: the conceptualization of
change of motion was a heroic achievement by the giants, from Galileo on,
who set the modern world in action.

Partial differential equations remain the primary tool for representing a
great many phenomena. Representations are often thought of as models.
Philosophers of natural science, but also, say, of economics, have heavily
emphasized the role of models ever since the publication of Nancy
Cartwright's path-breaking *How the Laws of Physics Lie* (1983). She was
the first philosopher of science to emphasize the extent to which physicists
conduct their work using a terminology of models, even if talk of models
evaporates when physicists begin to philosophize. (But we should not forget
Mary Hesse's *Model and Analogies in Science*, 1966.)

One reason I use the verb 'representing' rather than 'modelling' in
connection with mathematics is that 'models' often connote physical models, as in 'model airplanes' sold by toy manufactures. Material models, that
is, not intellectual ones. In aerodynamics, to which we turn in §27, there are
mathematical models of lift and drag, and material models of wings and
propellers tested in wind tunnels. Analogue models, as it has become
customary to say. (That's a bit of a pleonasm, since models *are* analogies!)
I shall keep the word 'model' for physical realizations and the verb 'to
represent' for the mathematics.

The representational–deductive picture is, as Manders said, deduction all
the way down. It is 'just proofs and more proofs' (§1.9), with representation
at the beginning and de-representation at the end. The picture can be
glimpsed in Francis Bacon's conception of mixed mathematics, which
'have for their subject some axioms and parts of natural philosophy' (§3
above). The picture is by no means restricted to the philosophers. It is pretty
much the description favoured in Gowers' *Very Short Introduction* to
mathematics (2002).

The picture makes clear sense of the attitudes I attributed to Mark Steiner
and Hannes Leitgeb, that applied maths is not maths. The representation
and de-representation at the ends are not mathematical but empirical; in

between there is just mathematics, albeit motivated by application. The trouble is that this picture is very much an idealization of the practice of applied mathematics. It may often make sense locally, but in the large it quite misses the interplay between the mathematics and the phenomena.

22 Articulation

T. S. Kuhn's *Structure* (1962) is written from a perspective very different from that of 'applied' mathematics. It was written at a time when philosophers of the sciences were still oriented towards theory rather than experiment, so all the action in the sciences takes place in a world of theorizing. But theories do not arrive neatly packaged and catalogued. A theory expressed using mathematical concepts needs to be fleshed out and to have its consequences laid bare. This is not a simple matter of deducing its consequences. It also involves clarification of concepts, removal of ambiguity, the making explicit of supplementary hypotheses, and much else. Kuhn called this 'articulation', and wrote of theory articulation (1962: 18, 97) and of paradigm articulation (1962: 29–36). He also spoke of experiment and theory being articulated together (1962: 61).

Although it is a bit protean, articulation is one of the most useful metaphors in Kuhn's incredibly rich book. In the case of theories, articulation is 'the process of bringing out what is implicit in the theory, often by mathematical analysis' (Hacking 2012c: xvi). But as the metaphor makes clear, I think, this is not just a matter of deriving mathematical consequences; it is also a matter of constantly remoulding and reorganizing the theory so that one *can* derive ever more precise consequences.

The representational–deductive picture completely passes by what is arguably one of the most important parts of theoretical research – what Kuhn called articulation.

23 Moving from domain to domain

One essential point about 'application' is that the same mathematical structures – models – often turn out to be useful representations of seemingly unrelated phenomena or fields of experience. Thus André Lichnerowicz, to whom I give the last word at the end of Chapter 7: 'We have gained the ability to construct models, the ability to say that while engaged in hydraulics we can say exactly the same thing we do in electrostatics, because we encounter the same equations in both fields' (CL&S 2001: 25).

Do we 'encounter', as if the equations are waiting for us? Or is this partly a selection effect? Once we are familiar with a model we try it out somewhere else. We economize. A body of mathematical practice developed for one domain is exported wholesale to another domain with which there are some similarities. It is not so clear whether we are discovering that the second domain has the same structure as the first domain, or whether we are sculpting the second domain so that it comes out shaped like the first. Probably both sorts of things happen. At any rate, the phenomena do not bear their equations on their faces, waiting to be 'encountered'.

In §§4.10–11, we noticed two prime examples of repeated use of a modelling strategy, while moving from domain to domain: polyhedra and harmonic motion. A single long paragraph of T. S. Kuhn's essay of around 1970, 'Second thoughts on paradigms', gives a brief, vivid account of moving from domain to domain (Kuhn 1977: 305f). The period goes from Galileo on the inclined plane, through Huygens on the pendulum, to Daniel Bernoulli's hydrodynamics (1738). That was an epoch when the mathematics and the physics were not distinguished. This work would, however, surely have been filed with mixed mathematics. But notice that it did not *yet* have any obvious practical uses.

Galileo observed that 'a ball rolling down an incline acquires just enough velocity to return to the same vertical height on a second incline of any slope, and he learned to see that experimental situation as like the pendulum with a point mass for a bob' (Kuhn 1977: 305). Huygens then transferred the idea of centres of mass. Much later Daniel Bernoulli 'discovered how to make the flow of water from an orifice in a storage tank resemble Huygens' pendulum' (1977: 306). Kuhn provides the details in the words of a physicist, whereas another author might tell them in terms of mathematical models, starting from Newtonian principles. But as Kuhn observes, the actual work at the time had 'no aid from Newton's laws'.

The representational–deductive picture is simple and clear. Yet Kuhn's examples make plain you don't have to use it at all when describing past events. *He* did not see things in the representational–deductive way.

As usual I would like to complicate things a bit. I shall use just three examples of 'applied mathematics'. One is in the rather long term: rigidity. The second is more localized: the wing of an aircraft. Finally, the most general kind of example: 'pure' geometry turning up all over the place in physics. The rigidity example illustrates the interaction between pure mathematics and applications. The flying example is a great instance of mission-oriented mathematics, replete with the idiosyncrasies and irrelevant details characteristic of true stories.

24   Rigidity

Nothing, it seems, could be more practical than rigidity. As soon as humans began to build shelters, they wanted structures that would not fall down. Some predecessor concept of rigidity must have emerged very early in human consciousness, not as a cognitive universal but as an ecological one. Take only a very late civilization about which we actually know something. The two classic structures of the North American prairie, where building materials were scarce, are the wigwam and the teepee. The former is a fairly permanent structure in which hide is fixed around a dome frame, while the latter is built around a movable frame of poles that becomes rigid when erected.

Well, actually rigidity is not as practical as it might seem at first sight. Harsh experience teaches that too rigid a building is bad news in an earthquake. The astonishing 101 building in Taipei is 101 stories tall. It rests in a major earthquake zone, which is also regularly beset by calamitous typhoons. (A catastrophic storm caused by an anticyclone in the Caribbean is called a hurricane; a similar phenomenon in the China Sea is a typhoon.) So 101 is built to sway. But not too much: it has a gigantic pendulum hanging inside the top floors, which will stabilize the whole in a time of shock. What we want is near-rigidity, with just enough flexibility to withstand storms and other upheavals. Wooden buildings stand up well, as evidenced by the quake in New Zealand in September 2010. It did not matter to the inhabitants of the Great Plains, but both tepees and wigwams (and, e.g., the yurts of the Gobi) are pretty quake-proof (but are destroyed by a tornado). They are 'nearly' rigid, which is what serves us better than rigidity.

It follows that there are two kinds of question to ask. What structures are rigid? But also, which are near-rigid, and will best withstand unusual stresses, such as earthquakes or hurricanes? The former question fairly quickly leads to abstract, non-empirical questions – pure mathematics – while the latter is eminently applied. Here we notice only a few facts about abstract rigidity, to illustrate the interplay between pure and applied.

25   Maxwell and Buckminster Fuller

In the early 1860s James Clerk Maxwell (1831–79) was lecturing at King's College London. In his classes he developed a number of 'geometrical' results with manifest applications to rigidity. In particular he determined the number of lines (e.g. bars of metal, or girders) 'needed to construct a stable figure joining a given number of points in space' (Everitt 1975: 170).

In general, n points demand $3n-6$ lines. This result became standard in British textbooks of civil engineering. But, writes Everitt,* 'there are exceptions', and 'Maxwell's original statement of his theorem embodies exactly the necessary qualification to cover' the anomalies. But Maxwell did not foresee how interesting the anomalies are. That was left for an enterprising engineer and visionary, none other than Buckminster Fuller (1895–1983).

The geodesic dome, or bucky dome, as it is often called, is one of the loveliest rigid structures. Fuller patented it in 1954 on the basis of hands-on work, not mathematics. The simplest example consists of six intersecting pentagons; the apex of each joins the next one at its mid-point. That makes for thirty points connected by sixty lines, thereby a counter-example to Maxwell's theorem unqualified.

Another way to think of a bucky dome is as a network of great circles that intersect to form triangles. Much the same effect seems to have been achieved by wigwams built by the Apache. The first such dome in an industrial society seems to have been designed around 1920 by the chief engineer of the Zeiss optical company: it was to host a planetarium. Fuller named all such spheres 'tensegrity spheres' and thereby created a science of tensegrity, short for 'tensional integrity'. Fuller enriched the concept of rigidity by tensegrity based on a balance between tension and compression, which has ample engineering consequences both in the large and the small.

Fuller intended his domes to revolutionize architecture – he even dreamt of covering an entire city in a dome. (Not a good idea, because materials needed to withstand the stresses are not practicable.) For ordinary habitation, a dome is a bad idea unless you totally change living habits – ordinary industrial furniture, all of which is built on the principle of the box, does not fit economically into a curved domain. (The Apache with their wigwams had another mode of life, and furnished their abodes quite differently.) Worse, bucky domes tend to leak at the corners. There is, however, a popular line in lightweight geodesic tents for campers, in which a tensegrity frame supports a continuous sheet of tenting material designed to cover it.

At present the largest structure designed using tensegrity principles is the Kurilpa Bridge for cyclists and walkers, over the Brisbane River in Queensland, Australia (opened in October 2009). The smallest human-built structures using Fuller's principles are named fullerenes. Spherical fullerenes, called bucky balls, are carbon molecules: allotropic forms of carbon. The first such to be made (in 1985) was C_{60}. By 1992 geologists had detected natural fullerenes on the surface of the earth. An 'artificial' kind of substance turned out to exist 'in nature' as well, but people do it better.

Carbon nanotubes are generalizations of spherical fullerenes, and the toughest stuff yet made.

Hence it surprisingly turns out that the greatest interest in bucky-structures is not in human-sized architecture, but in mesophysics (between micro and macro) – the domain, for example, of carbon nanotubes. That is all engineering, however large or small. But not micro: we are not yet at the quantum level. It is applied science, not even applied mathematics – but tensegrity has generated its own field of what can only be called pure mathematics.

We shall turn to that in a moment, but first another instance of Fuller's imagination. He seems to have thought that the geometry of the universe is constructed out of densely packed tetrahedrons, an argument based on some results about close-packing of spheres and global requirements for the stability of literally everything. Plato's *Timaeus* vindicated! Maybe Fuller was visiting a town that sold milk in Tetra Paks, the tetrahedral milk containers invented in Sweden, first marketed in 1951, and quickly accepted in Europe and then in many parts of the world. (I first used them in Uganda in the 1960s.) The tetrahedron is a particularly efficient kind of packaging, minimizing surface area to hold a half-litre of milk. You can also pack far more containers into a given volume than with many other shapes. What is the best way of doing that? How 'efficiently' can you pack tetrahedrons, so as to minimize the amount of empty space? This is a question in the calculus of variations.

Part of Hilbert's eighteenth problem is about packing spheres: a conjecture named after Kepler, who appears to have proposed that the best packing is cubic or hexagonal. This conjecture was confirmed (and probably proved) by computer exhaustion of cases over a decade ago. In passing, Hilbert mentioned the problem of packing tetrahedrons. People who like puzzles can experiment with conveniently available little plastic objects, namely pieces from *Dungeons and Dragons*. Sophisticates use computer simulations.

Maybe this is just 'recreational mathematics', but it really took off as a sort of subfield in 2006 when John Conway attacked it. (He is the truly innovative mathematician, but also puzzle fiend, encountered in connection with the game *Life* in §2.28, and as the man who named the Monster and shouted 'Moonshine!' in §3.10.) He got a packing efficiency of 72 per cent. That was too low! There have been a whole series of improvements published in the likes of *Nature* and *Science*, with a current best of 85.63 per cent set in 2010.

Incidentally, the general bin-packing problem of this sort is not known to have a polynomial algorithm, although interesting cases are solvable. But the general question is, in a technical sense, at least as hard as the hardest NP problem, and might even be a counter-example to P=NP. Here we have a not so untypical mix of different types of exploration (as I shall call it), some using material objects to see how they can be arranged, others simulating by computer various types of arrangement. They begin from varying interests, some practical, some playful. Then we are led to a really deep problem in pure mathematics.

Is this pure mathematics or applied? Is it play or deep mathematics? Were Plato and Buckminster Fuller serious in speculating about the structure of the universe while Conway and others were just playing? Such questions begin to lose their force.

26 The maths of rigidity

There is a wonderful mathematics of rigidity. Some say it can be read into Euclid, but only in 1814 did Cauchy, building on ideas of Lagrange, prove the foundational theorem. Any convex polyhedron with rigid plates, but with hinges at its edges, is, despite the hinged edges, rigid. (Strictly speaking, Cauchy 'almost' proved this; tidying up the lemmas took more than a century.) Here we move from a seriously practical problem (what stays up?) to 'pure' mathematics. An entire discipline, which becomes an aspect of topology, develops. But then real life strikes back, and the mathematicians learn from the engineers.

Late in the nineteenth century a French engineer, Raoul Bricard, had discovered flexible non-convex polyhedra. For of course Cauchy had convex structures in mind. The story in many ways replicates Lakatos' *Proofs and Refutations* (1976), with its account of star polyhedra – the most readily envisaged non-convex polyhedra – providing what Lakatos called 'monsters' for a theorem of Euler. Some tried to insist they were not 'real' polyhedra, or at least not what the theorem 'meant'. Lakatos' moral was that monster-barring is bad methodology, for it inhibits new concepts and deeper mathematics. The flexible polyhedra can be thought of as 'monsters' for Cauchy, but their physical realization is perhaps more interesting than Lakatos' star polyhedra. You can do something with them – flex them!

The discipline now named *rigidity theory* is said to have begun in 1960 with a series of theorems proved by Atle Selberg. He was mentioned in §1.11 for his proof of the prime number theorem. It was a truly pure mathematician who created rigidity theory as a separable discipline! Indeed he has been called the mathematicians' mathematician, the purest of the pure.

Robert Connelly has been the leading tensegrity mathematician. For a brief explanation of the many types of stable structures that have emerged in the past half-century, see Connelly (1999). In 1977 he began to elaborate a whole class of flexible polyhedra. Does this have practical consequences? Not obviously. No one is going to build a flexible polyhedron as a dwelling or for any other foreseeable purpose. But there is an 'unintended application', *App 6*. You can make a flexible polyhedron for an exhibition, to create a hands-on exploration for children of all ages. There is one at the National Museum of American History in Washington. (That sounds like the wrong place to store it, but that museum began as the Museum of History and Technology, and was renamed.)

Flexible polyhedra are known in 4-space. Are there any in more dimensions than that? That is a question investigated, at present, for purely 'aesthetic' reasons; the answers are not known. Pure – for the nonce.

Lest you think we have now got away from the deep physics of the universe, and Wigner's unreasonable effectiveness, I should mention that Connelly's work includes percolation theory. Now that sounds like the most applied subject you could imagine. Given a slice of porous material, what are the chances that a liquid poured on top will make it to the bottom? This leads to the theory of random groups of points, edges, and vertices.

That is a splendid example of modelling. Real substances – ground coffee or, more interestingly, layers of limestone beneath the surface of the earth – are a mess. So we make very simple representations that try to capture salient properties, and this in turn generates really hard mathematical problems. The representation-deductive picture seems on the right track here.

Polyhedra are characterized by their points, edges, and vertices. You could think of percolation theory as the statistical theory of polyhedra. Imagine a stratum built up by chucking in, at random, all sorts of polyhedra, and consider how drops of fluid would percolate through it. When well modelled, this becomes a 'pure' investigation. But then it is incorporated in the statistical mechanics of electromagnetic interactions. This in turn becomes a classic case of Wigner's unreasonable effectiveness: it leads on to the central use of renormalization in quantum field theory. That in turn generates pure mathematics and to the Fields Medal awarded to Stanislav Smirnov in 2010. Smirnov's work is embedded in the Langlands programme (§1.13), which in the 1980s took in exactly those groups of points, edges, and vertices that went into the Ising model in statistical lattice theory that then got applied in renormalization theory.

Perhaps one could rewrite the practice and history of rigidity to fit the representational–deductive picture, but it would be an irrational reconstruction. The picture is attractive in small segments, but when we look at the larger picture, there is too much cross-fertilization between the phenomena and the mathematics to make it ring true. Applied mathematics is seldom 'deduction all the way down' between representation and de-representation. Correspondingly, the Leitgeb–Steiner attitude, that there is (pure) mathematics on the one side, and applications on the other, seems also to be inadequate.

27 Aerodynamics

Lift and *drag*: These two four-letter words, which (like most of the formerly unprintable four-letter words of English) are both verbs and nouns, point at what we want to know about the wing of an aircraft. Lift, crudely, is what gets the airplane into the air and defies gravity; drag, crudely, is the force that opposes motion of the body through the air. The first aircraft were designed by trial and error. Applied mathematics then helped engineers make better and better wings. David Bloor's *The Enigma of the Aerofoil* (2011) is a marvellously detailed study of the design of wings in Germany and Britain, 1909–30.

Bloor is well known as a co-founder of the Edinburgh Strong Programme in the Sociology of Scientific Knowledge, which has had such an effect on the development of the discipline now called science studies. As is to be expected, his book relates technoscience in action to its institutional set-tings. His book is an outstanding (and mature) example of this genre. My use of it hardly touches on its many excellences. Even so, what I use will seem too local to bear on philosophy, for it is about a very specific moment and two very specific locales. As will be obvious by now, I believe that hard examples have much to teach about facile generalizations.

First: what is the 'enigma'? The title of a lecture that Bloor gave in 2000 (mentioned in Eckert 2006: 259) explains. 'Why did Britain fight the First World War with the wrong theory of the aerofoil?' Good question, to which the book is one long answer.

Second: what is an aerofoil? The word was invented by F. W. Lanchester (1869–1946) in 1907 to mean a structure that gives rise to a lift force when moving through the air. He wanted it as a contrast to the word 'aeroplane', which in addition to its present meaning, a flying machine with fixed wings, had also been used to mean just the wing – a wing being a *plane* surface that moved through the air.

Bloor's enigma is not about aerofoils but about their history. Lanchester sketched what has become the accepted picture of lift, but his research programme was completely rejected by his own (British) community in favour of a very different approach that degenerated and had to be abandoned. His programme was, however, developed to fruition in Germany. Why?

Bloor's answer analyses two different types of institution, with two very different conceptions of the relation of mathematics to its applications. This confirms that the boundaries between pure and applied are by no means in the nature of the subject matter. In the context of the present chapter on applied mathematics, that is the chief use that I shall make of the book. But of course this point is incidental to Bloor's rich history.

The Strong Programme advocated a thesis of symmetry. Historical analysis should give coherent explanations of the use and development of scientific theories, regardless of whether the theories are later deemed to be true or false. Historical explanation should be 'symmetric'. This doctrine is to some extent a recasting of older historiographical advice to avoid anachronism, and not compose whig history (as the conservative historian Herbert Butterfield called it) – whig history (exemplified by David Hume's *History of England*) encouraging the idea that what happened was right, the past preparing the space for what is now the status quo. Bloor's book is almost of necessity symmetric in that it recounts non-judgementally what happened in the rival research programmes.

History can be told from complementary points of view. Michael Eckert's *The Dawn of Fluid Dynamics* (2006) is a magisterial account of the German history of early aeronautical engineering. The two books focus on the same figure in Germany, namely Ludwig Prandtl (1875–1953), but Eckert's account is necessarily richer in German detail than Bloor's. Moreover, Eckert is an invaluable guide to one story of interactions of technology, science, and mathematics. His subtitle is, *A Discipline Between Science and Technology*. Bloor's methodology is less inclined than Eckert's to distinguish science from technology.

28 Rivalry

Bloor writes about the very first technical problem in aviation. What makes the wing of an aircraft lift? What is the best shape for taking off, flying, and landing, when the plane has an engine of a certain amount of power? Given that the best-remembered early airplanes were mostly American and French, it may come as some surprise that Bloor has chosen the rival theories of German and British scientists. The early achievements now recalled are the American Wright brothers, and, to a lesser extent, the

1909 flight of Louis Blériot across the English Channel. But it was the Germans and the British, both preparing for war, who invested public monies in research. (Indicatively, *Webster's* dates the American spelling 'airfoil' as first used in 1922, although 'airplane' came into American usage in 1907.)

Blériot's flight scared the Brits into action: they were no longer on an island made impregnable by the sea and the Royal Navy. (Blériot peacefully financed his airplanes and his famous flight from the profits of his automobile headlamp, which dominated the early market.) Preparing for war in 1909? Great Britain and Germany were engaged in a grossly expensive naval arms race, and flying changed the rules.

The rivalry in the air was not just about who spends most. Bloor's subtitle is *Rival Theories in Aerodynamics 1909–1930*. His topic is less rival theories than rivalry between two different traditions of research. The traditions are, as always, embedded in two different sets of institutional arrangements and histories. One of these, British, gave precedence and status to the mathematicians. The makers and shakers were products of Cambridge University, whereas Lanchester was 'just' an engineer, whose ideas about flight were simplistic and to be disregarded. The other tradition, German, paid more attention to the engineers, and *its* leading mover and shaker, Ludwig Prandtl, developed Lanchester's picture successfully and designed the better wings, the better aerofoils. This rides well with the familiar truism that the engineer wants a useful answer, one that leads to making something that works adequately, while the mathematician wants an answer that is demonstrably right, and which explains why it is right. (One caution: Bloor makes Lanchester more significant in the German story than Eckert does.)

This difference in aim introduces a word I have thus far omitted: idealization. You might think that it is the mathematicians who want the idealizations, and the engineers who want the messy real world. Bloor's story shows how misleading that is.

We are concerned with motion of a solid object in a fluid. At the time of, say, Daniel Bernoulli, author of *Hydrodynamica* (1738; §21 above), water was the manageable fluid. But the air was treated as a fluid too. Indeed Bernoulli's approach radically enriched the kinetic theory of gases. Equations developed by him and Euler (1707–83), to some extent after collaboration, gave a rich analysis of motion in an ideal gas. Here the idealization was of a uniform friction-free (non-viscous) fluid. But of course turbulence rules in real life.

Only in the nineteenth century did a good understanding of dynamics of real fluids begin to emerge. The basic equations – yes, souped-up partial differential equations – that still stand were developed by the mathematician who dominated Cambridge applied mathematics, George Stokes. Of course there were other even more luminous figures, such as Maxwell, but Stokes' work was exemplary. He worked out the equations that are still at the core of fluid dynamics.

Claude-Louis Navier was a brilliant French engineer in an old-fashioned sense – he built bridges. He also generalized the work of Euler and Bernoulli to allow for inter-molecular attraction, thus providing a representation of viscous liquids. His model made no allowance for many factors later recognized as important, some of which were captured by Stokes, but formally his equations are essentially those that Stokes developed. Hence they are now called Navier–Stokes equations, but for brevity I will call them Stokes'. We still do not know if these equations always have smooth solutions: that is indeed one of the seven open Millennium problems (§2.21).

Do the mathematicians want the idealization and the engineers the real world? In a sense Bloor's analysis runs the opposite way – in this case. The mathematicians (to grossly oversimplify) wanted analyses of the aerofoil that relied on the most realistic available modelling: that is, as special solutions to Stokes' equations. The engineers (oversimplifying again), starting with Lanchester, were content with manageable calculations, basically working with the old Bernoulli–Euler picture. Thus they were content with radical idealization, while the mathematicians wanted a deeper, more realistic analysis.

29   The British institutional setting

At the beginning of the twentieth century, Cambridge mathematicians and their pupils thought that idealized non-viscous gas was of no interest whatsoever. All the problems had been solved. Viscosity and turbulence are where the action is. They could not solve the Stokes equations in full generality. (Nor can we today.) So the task must be to provide solutions that bear on the case of a wing-shaped object moving at such and such velocities thorough real air. We want optimum solutions for the shape, for the angle, the placement of the wing or wings.

Lanchester, I have said, was an engineer. While employed in various engineering works he developed a whole series of innovations and patents. He made enormous improvements in the petrol engine, and in 1899 established a company for making motor cars. It was poorly managed, but its products were on the cutting edge of technology. As soon as manned flight was established, he devoted much effort to designs that improved lift and its control. His picture of what happened was based on a fluid continuum modelled on the idea of a friction-free and uniform batch of infinitesimal fluid elements which exerted force on a moving plane, the aerofoil. To the mathematicians, this picture seemed to be obsolete. Real gases were not like that. So the British mathematical establishment, co-opted by the government in various committees and working groups, dismissed Lanchester's proposals.

Bloor heads his first chapter with an amazing sentence written down in 1916. It could serve for self-parody. 'In the meantime, every aeroplane is to be regarded as a collection of unsolved mathematical problems; and it would have been quite easy for these problems to have been solved years ago, before the first aeroplane flew' (Bloor 2011: 9). Deduction all the way down, with a vengeance!

Bloor describes the man who penned that sentence as 'a brilliant and opinionated mathematician', and a wrangler. That is, he came in the first class in the Cambridge undergraduate mathematics examinations, the 'Tripos'. Bloor often refers to the British tradition in mathematical aeronautics as Tripos-oriented.

The Tripos was not just an examination, but a training ground in a local way of doing mathematics, which lasted most of its wranglers the rest of their lives. J. E. Littlewood, source of my fourth epigram, has a whole section on 'interesting' Tripos questions in his *Mathematician's Miscellany* (§1.28). It is just possible that anglophone philosophy of mathematics owes its representational–deductive picture of applied mathematics to the Tripos tradition to a far greater extent than is commonly acknowledged. Hence the tradition may have done untold and unpredictable harm, just as, in my opinion (and that of many mathematicians of my acquaintance), G. H. Hardy's fantasy about pure mathematics, *A Mathematician's Apology*, did much harm to philosophers.

To get an idea of the Tripos, many of the exercises in Hardy's wonderful textbook, *A Course of Pure Mathematics* (1908) are stated with the year of the Tripos in which they were set; the Preface to the seventh edition (1934) states, 'I have inserted a large number of new examples from the papers for the Mathematical Tripos during the last twenty years.' Those were all questions about what Hardy judged to be pure mathematics, but there were also plenty from what Cambridge judged to be applied mathematics – in the style of Stokes. Every Tripos included what were in effect research questions that in a few hours could sort the wranglers from the rest. The British establishment tackled the aerofoil as if it were a collection of very hard Tripos questions. Hence it looked condescendingly on a mere auto-motive engineer.

None of this is intended to imply that all this work was a priori. On the contrary, the mathematical representations were constantly being tested and revised using experiments in wind tunnels. Early twentieth-century British wind tunnel technology was for long the best in the world. This required a great deal of new instrumentation in order to determine wind velocities, points of turbulence, and much more. But even when a theoretical analysis appears, in retrospect, to have been definitively refuted, the observed results were, as Bloor shows, reinterpreted so as to require mere revisions. In fact 'falsification' was always closer to hand precisely because the British research programme attended to first-class data about turbulence. The rival research programme in Germany did not have access to so many empirical facts, and so could get on with designing the aerofoil. The relations between purity and application are often the reverse of what the naïve philosopher would expect!

30   The German institutional setting

Bloor reasonably presents the matter in terms of engineers and (or often versus) physicists, but also 'practical men' and mathematicians. In the British case, the theorists were inextricably bound up with a mathematical training at Cambridge. In the German case, the *technische Hochschulen* (what in English would have been called polytechnics) were the site of much of the research. A leading figure in the story is Ludwig Prandtl, who according to Bloor was more sympathetic to Lanchester's engineering than were the British mathematicians. Prandtl's career began in the Munich Polytechnic; he then moved to the Hamburg Polytechnic. From there, a paper of his was so admired that he was called to Göttingen.

But not the mathematical Göttingen so familiar to philosophers and historians of mathematics. It is not the Göttingen of David Hilbert, but that of the institutionally powerful Felix Klein (1849–1925). Certainly the latter was a great mathematician, but also one keen to integrate all of mathe-matics, pure and applied. He actively sought out engagement with industry, collaborating, for example, with the chairman of Bayer, the chemical and pharmaceutical giant. Together they established a Society for Research in Applied Physics and Mathematics, which in turn helped establish a uni-versity department of Applied Mathematics and Mechanics in 1905, with Prandtl as one of the two professors. Klein urged his appointment as one 'who combined the expert knowledge of the engineer with a mastery of the apparatus of mathematics' (Bloor 2011: 256).

It should be said that many Göttingen mathematicians were just as hostile to this import from the lower-status *Hochschulen* as the Cambridge applied mathematics fold was to 'practical men'. But German paymasters wanted to support, and learn from, the contributions of the engineers. Throughout the twentieth century Britain notoriously suffered from inept relationships between science and industry. They seldom troubled German enterprise. But let us turn from these merely incidental points to something more conceptual.

31 Mechanics

Bloor's important Chapter 6, '*Technishe Mechanik* in action', makes us realize that the very notion of applied mathematics, which I have been taking too much for granted as merely the counterpart of the pure, may have different connotations in different traditions. Why wasn't Mechanics included under Applied Mathematics at Göttingen, with no need to say, 'and Mechanics'? This question (of interest only in a philosophical chapter on applied mathematics!) prompts a few words on mechanics.

'The idea of trickery – and, ultimately, of violence – appears in the word "mechanics", since *mēkhanē* signifies "trick"' (Hadot 2006: 101). Pierre Hadot, the great scholar of Neoplatonism, was emphasizing that the scientific attitude derives from Prometheus, who stole fire from the gods by trickery. The *OED* teaches that in English the meaning of 'mechanics' is (a) 'originally (and still in popular usage): That body of theoretical and practical knowledge which is concerned with the invention and construction of machines. (b) The department of applied mathematics which treats of motion.' Mechanics, in English usage, *does* fall under applied mathematics.

The word *Mechanik* entered German much later than English, and has a distinct connotation in connection with applied science. The Göttingen Institute was for Applied Science *and* Mechanics. In 1935 an English colleague told Prandtl that he ought to have won a physics Nobel Prize for his work on aerodynamics. Bloor (2011: 194) quotes Prandtl's reply: 'At least according to the division of the sciences that is usual in Germany today, mechanics is no longer considered to be part of physics. Rather, it stands as an independent area between mathematics and the engineering sciences.'

Bloor goes on to argue that Lanchester's approach to the analysis of lift and drag was eminently acceptable in such an environment, and was therefore taken up and made central to the field. But on Lanchester's English home turf, the mathematicians wanted a 'deeper' account of lift that paid serious attention to viscosity and turbulence. Or, to put it in a different way, Lanchester's conceptual home turf was mechanics as under-stood in Germany, rather than applied mathematics as understood in England. This is part of Bloor's solution to his enigma.

32   Geometry, 'pure' and 'applied'

We have mentioned Michael Atiyah a number of times, most recently in §4.10, enthusing about the ways in which the mathematics of polyhedra recurs in the natural sciences. He began his career as what might be called a pure geometer; he made many major contributions to algebraic topology. Together with Isidore Singer he developed what is known as the index theorem in the 1960s: it connected pure mathematics in the form of differ-ential geometry with applied mathematics in the form of quantum field theory. The interchange had radical consequences for both fields.

A lovely end-of-career account of the reunion of 'pure' and 'applied' is to be found in a recent survey article, 'Geometry and physics' (Atiyah, Dijkgraaf, and Hitchin 2010). The doors between 'pure' geometry and physics have been unlocked, and mathematicians have had to learn how to think like physicists – and vice versa.

In one case the doors were literally unlocked. Trinity College Cambridge has a formal dinner to celebrate a Fellow's eightieth birthday. Atiyah's fell in 2009, and he gave the mandatory after-dinner talk. It seems that when he was at MIT, he discovered that a door separating mathematics and physics was permanently locked. Why? The physicists had just installed new carpeting and did not want the grubby mathematicians messing it up with muddy snow on their boots. (The talk was printed in the College's annual *Record* for 2009.)

33   A general moral

Atiyah's recollection is a parable. For historical reasons described in Part A of this chapter, pure and applied mathematics were distinguished and then, in many influential places, kept pretty separate. Perhaps this encouraged the philosopher's 'representational–deductive' picture, where the mathematics went on in the deductive part of reasoning, while the applications were a matter of representing experimental or observed phenomena by mathematical formulae or structures. The doors between the two, pure and applied, were opened only at the start of a deduction and at its end. I hope that even the very few examples given in this chapter will have shown that is not, unequivocally, how it is. It is a useful picture of mathematical activity, *sometimes*.

I hope that my various examples illustrate that the very idea of 'applied mathematics' is an admixture of different pictures of intellectual and practical activity, and that these can be classified in different ways in different traditions. Some ways jibe better with the representational–deductive picture, for example, the 'Tripos-oriented' programme described by Bloor. Others do not conform at all to that picture, for example the 'theoretical mechanics' also described by Bloor. When we get down to details, my classification of *Apps* looks less and less robust!

34 Another style of scientific reasoning

Already Kuhn's idea of the articulation of theories by, among other things, mathematical analysis, illustrates the fact that activities, which can be described as applied mathematics, can be described in many ways. Since it is important not to be beguiled by a single mode of description, I shall conclude this chapter with another mode, parallel to the way in which I ended Chapter 4.

I introduced a seemingly quite independent line of research, my 'styles project', in §4.22. Following A. C. Crombie I find that a small number of fundamentally different genres of inquiry have emerged in the history of the sciences in Europe – ways of finding out that have later become global. The third of these 'styles', in Crombie's template, was what I call, for short, hypothetical modelling: in Crombie's original form of words, 'the hypothetical construction of analogical models as an explicit method of scientific inquiry' (Crombie 1994: 1, 431). Galileo, in my view, 'mathematized' the world. This way of looking at things is heavily influenced by Husserl's Galilean chapters in his 1936 *Crisis* (Husserl 1970). Of course Galileo also participated in the radical transformation of Crombie's second style of thinking and doing in the European tradition, namely experimental exploration.

Crombie describes Galileo's predecessors as making analogical models. After Galileo, the models became increasingly mathematical rather than material or pictorial. I argue that the making of hypotheses as models of some aspect of the world, and casting them into mathematical form, became one of the fundamental styles of scientific reasoning in the European tradition (third, in Crombie's enumeration, after mathematical demonstration and experimental exploration). This opinion superficially jibes with the representational–deductive picture presented in §21 above. Indeed someone could schematize even further. (a) We make a hypothetical model of a phenomenon in mathematical form. That is hypothetical modelling. (b) We articulate the mathematical properties of the model by deduction.

(c) We test the conclusions of (b) against the world experimentally. This tidy scheme (a)–(c), which happily no one has promoted in exactly this form, would be a paradigm of how not to practise philosophy. In contrast we should think of the constant interplay of different genres of inquiry in scientific activity. The examples presented in §§24–32 illustrate the ways in which these styles fruitfully interact.

CHAPTER 6

In Plato's name

1 Hauntology

Philosophizing about mathematics is haunted by platonism, both totally naïve and enormously sophisticated. It is supposed to be a kind of ontology, but one is tempted to recycle Jacques Derrida's (1994) brilliant pun and call it *hauntology*. I had better explain. In *Specters of Marx*, based on a pair of lectures given in 1993, Derrida plays effectively on the repeated occurrence of ghost-words in Marx and Marxism. He starts with the first words of the preamble to the *Manifesto*: 'A spectre is haunting Europe – the spectre of Communism'. Derrida's message is very serious, the claim that hauntology is more embracing than ontology. Marx, who fights ghosts, is contrasted to Hamlet, who succumbs to them.

Once in place, the play on words is readily recycled. The British music critic Simon Reynolds uses 'hauntological' to describe UK electronic music that is created primarily by artists who manipulate samples culled from the past (mostly old wax-cylinder recordings, classical records, library music, or postwar popular music) to invoke either a euphoric or an unsettling view of an imagined future. He calls it 'an uneasy mixture of the ancient and the modern'. The very same words seem apt when applied to contemporary platonism in the philosophy of mathematics.

No ghost more effectively haunts all Western philosophy than Plato's. Hauntology is an idea that fits a lot of scenes, but aside from Marx, who invoked spectres often, none fits it better than Plato and ontology. A scary site looms up through the fog of midnight, a graveyard haunted by the spectres of Plato and Marx. Many have posed as exorcists, but none have succeeded.

2 Platonism

Plato's words have launched a thousand doctrines, many named after him. Many of those, one may feel, take his name in vain. That is true even within a discourse as parochial as the philosophy of mathematics. My capitalization convention, announced in §1.18, does not help all that much. It is handy to keep lower-case *p* platonism for recent ideas that often have a highly semantic component, and to keep the capital *P* for what is plausibly connected to the historical Plato – but what is the historical Plato? Curiously, hardly anyone seems to challenge the use of his name.

One philosophical logician who does so is William Tait* (2001). I find his own delineations of doctrines a bit hard to grasp, but his labels are suggestive. 'The view that there is an external criterion of mathematical existence and truth and that numbers, functions, sets, etc., satisfy it, is often called "Platonism", but Plato deserves a better fate.' Tait calls the usual so-called Platonism *super-realism*, and I think he is referring to, among other things, the idea that mathematics is just 'out there' to be explored – and, moreover, that it is populated by mathematical 'objects'. What fate *does* Plato deserve, in his opinion? A more modest doctrine that Tait calls *realism*, and which he describes as 'the default position'. It looks as if part of the default position is the oft-quoted maxim that Michael Dummett (1978: xxxviii) attributes to Georg Kreisel's review of Wittgenstein's *RFM*: 'the problem is not the existence of mathematical objects, but the objectivity of mathematical statements'. What Kreisel (1958: 138 n. 1) actually wrote is this: 'Incidentally, Wittgenstein argues against a notion of mathematical object (presumably, substance), but at least in places (IV, §35, 243; III, §71, 197; or III, §50, 250f), not against the objectivity of mathematics, especially through his recognition of formal facts.' (I have changed Kreisel's section numbers from Wittgenstein 1956 to Wittgenstein 1978.)

Tait's 'default position' is both sensible and attractive, but, so far as I can see, it has no textual connection with the historical Plato. Nor is it the present default meaning of 'platonism' – which, in recent history, is aptly captured by the label 'super-realism', and is much concerned with ontology and the existence (or not) of mathematical objects.

Tait thinks that Plato deserves a rather ho-hum realism. I like my philosophers bold, so I suggest that Plato 'deserves' – if that makes sense at all – *more* than super-realism. He deserves a Pythagorean crown, preposterous but daring. But that is not the present way of thinking.

In order to avoid quarrels over names, this chapter is built around three mathematicians who take the label 'Platonism' as self-evident, one avowing it, two disdaining it. The super-realism of the self-described Platonist is interesting because it holds that only a fragment of the mathematical universe is super-real or, as he puts it, primordial.

3   *Webster's*

Before proceeding to the mathematicians, what do ordinary people understand by Platonism, if they understand anything at all? Here, for the last time, I shall consult a dictionary, *Webster's Third New International*. Recall from §2.4 that it is still pretty much as published in 1961. Oddly, although it sets the American standard, it is almost the only dictionary that dates quickly, for it tried to be up-to-date fifty years ago and thus is dated. Yet it serves to remind us how much has been done, in different epochs, in Plato's name.

> **pla-to-nism** 1 *usually capitalized* : the philosophy of Plato stressing that ultimate reality consists of transcendent eternal universals which are the true objects of knowledge, that knowledge consists of reminiscence of these universals under the stimulus of sense perception, that objects of sense are not completely real but participate in the reality of the ideas, that man has a tripartite preexistent and immortal soul consisting of the appetitive functions, the spirited functions, and the intellect, and that the ideal state is aristocratic and made up of three classes of artisans, soldiers, and philosopher-rulers.
>
> 5 *sometimes capitalized* : a logical or mathematical theory incorporating within its language names for such abstract or higher level entities as classes – contrasted with *nominalism*.
>
> **nom-i-nal-ism** 1a . . . *specifically* a theory advanced by the medieval thinker Roscelinus that universal terms such as indicate genus or species and all general collective words or terms such as *animal, man, tree, air, city, nation, wagon* have no objective real existence corresponding to them but are mere words, names, or terms or mere vocal utterances and that only particular things and events exist.
>
> 1b : a logical or mathematical theory excluding from its language any names or variables for such abstract or higher-level entities as classes – contrasted with *platonism*.

4   **Born that way**

Samuel Taylor Coleridge (1835, 1: 192), poet, philosopher, and much else, opined that 'every man is born an Aristotelian or a Platonist'. There is a germ of truth in that; it makes it pointless to ask, for example, why my own instincts have always been crudely nominalist in an undisciplined mediaeval way, vaguely reminiscent of Ockham, and too crude to be called Aristotelian. I came that way. So, one may flippantly suggest, did several major figures, including Goodman, Quine, Russell, Mill, and Hobbes. A certain cast of mind and maybe stomach is noticeable in these great philosophers. It remains, even when nominalism seems formally untenable with other aspects of their philosophies. Thus Quine is now referred to as a platonist because of his argument that numbers and other mathematical objects are indispensable to contemporary sciences, and so the sciences presuppose the existence of numbers etc. I am inclined to call that the wan platonism of a free-born nominalist.

5   Sources

The Roscelinus in *Webster's* entry of nominalism is hardly a household name, not even in French: Roscelin de Compiègne (1050–1121). For many years his only text known to survive was a really nasty letter to his former pupil, Peter Abelard (1179–42). Roscelin maintained that the three Persons of the Trinity had to be three distinct things, said to be a consequence of his theory of names. He was best known in his day from Anselm's (1033–1109) mocking denunciation.

It is not good for one's reputation to be remembered (if at all) by what opponents of the calibre of Anselm and Abelard said about your ideas. Roscelin seems to have had to recant, on pain of being declared anathema, excommunicated, and stoned by the public; yet he did go on teaching his philosophy of language. Recent scholarship presents him as an interesting person and a coherent thinker (Mews 1992, 2002).

Obviously none of this has anything to do with mathematics, but the dictionary does remind us that nominalism could get you into far hotter water in days gone by than is the case today.

In the late 1950s one could have found many American philosophers who defined Platonism ('usually capitalized') as *Webster's* did in 1961, but I doubt that as many would do so today.

One imagines that a student of Quine was consulted for the definitions that bear on the philosophy of mathematics, platonism (definition 5), and its contrary, nominalism (definition 1b). Nominalism in the philosophy of mathematics has moved on since then, but most analytic philosophers of mathematics would say that *these* parts of the dictionary entry are on course. The pedantic among us note that platonism is not 'a mathematical theory incorporating names for abstract or higher level entities such as classes', but a philosophical theory about the names allowed in a mathematical theory; same emendation for the entry for nominalism. The dictionary probably meant that a mathematical theory incorporating names for abstract or higher-level entities such as classes is platonistic.

6   Semantic ascent

Platonism, however it be defined, is ancient, tied to Plato, whereas nominalism is a product of scholastic philosophy. Yes, Aristotle may have planted the seeds of anti-Platonic mediaeval nominalism, but those seeds germinated and then flourished only in the Middle Ages. *Webster's* was right about that.

All kinds of nominalism are in some way semantic, for as their name implies, they are about names. The good old-fashioned Platonism of Webster's definition (1) is about what there is, while the relatively novel platonism of Webster's definition (5) is a linguistic doctrine, about what words to use. Likewise nominalism (1b) is about which words *not* to use. The one welcomes names for 'abstract objects', including names for numbers, and the other excludes them, or analyses them away.

In Quine's splendid phrase, introduced in the final section 56 of *Word and Object* (Quine 1960: 270–6), platonism (5) is the result of *semantic ascent* on something only very loosely connected to Platonism (1). Many recent analytic philosophers might argue that platonism (5) is the only way to make sense of the hodgepodge of aspirations expressed by the definition of Platonism (1).

I try to avoid semantic ascent. Any time John says something, and Joan retorts that it's not true, Joan is, in a natural understanding, saying something about what John said – 'semantic ascent'. No problem. But if semantic ascent means turning thoughts about things into thoughts about words, I try to avoid it as much as possible. I am evidently much interested in how words are used, and how false generalization from one use to another may encourage philosophical conundrums. But my own philosophical practice is to try to stick to the 'object level', to the level at which we started, rather than moving up to debates about how we ought to talk. This is, perhaps unfortunately, not to subscribe to the exquisite defence of semantic ascent in the final section of *Word and Object*. It is so marvellously written, in Quine's unique style, that one almost regrets not subscribing.

It is worth repeating the difference between traditional nominalism and recent nominalism in the philosophy of mathematics. Mediaeval nominalism held that there was no universal, 'wagonhood', associated with the collective noun 'wagon'. Individual wagons exist, but wagonhood does not. A proclamation, such as 'Wagons shall be taxed at 14 per cent', says that each individual wagon shall be taxed at 14 per cent of the sale price. It is not, according to the mediaeval nominalist, saying something about the class of wagons, for there is no such class; there are only individual wagons.

Thus mediaeval nominalism can be understood, anachronistically, as a philosophical doctrine about the use of collective nouns, such as 'wagon', but not as prohibiting their use.

Contemporary nominalism moves up a level. Thou shalt not utter any names for abstract objects! Thou shalt not speak of numbers, for they are abstract, are they not? A resurrected (and reconstructed) nominalist from olden times might urge that, in mathematics, such words are not *used* as names for abstract objects at all. Thus there is no need to exclude them.

7 Organization

After these prolegomena, we proceed to the opposed views of two senior twenty-first-century mathematicians, Alain Connes and Timothy Gowers,* augmented by an older colleague of Connes, André Lichnerowicz, a mathematician with some roots in Bourbaki. Their ideas are best described as *attitudes* to the mathematical life. Thus they are opposed philosophies, in a popular (and honourable) sense of the word 'philosophy'.

These are not the technical doctrines of philosophers who dedicate their lives to careful analysis. They are the background ideas of men who dedicate their lives to creative mathematics. Their opposed philosophies may not matter much to how they engage in mathematical research, but they make a difference to how they conceive of their activities. They are philosophies to live by. Each mathematician's philosophy makes sense, for the individual, but what makes sense of the mathematical life for one person may be the opposite of what makes sense for another person.

The opposed philosophies of mathematics may, however, make a difference to the way in which mathematicians teach their subject – to the extent that Lichnerowicz was responsible for radical reform in the teaching of mathematics – what in America became known as 'the new math' (see §9 below).

It may seem strange, in a sequence of philosophical thoughts about mathematics, to spend so many pages on the rather casual (but heartfelt) reflections of a few mathematicians, and not to consider up-to-date philosophy of mathematics. We shall come to that piecemeal in Chapter 7, but even there the approach will be unusual. First we will consider how the noun 'platonism' entered the philosophy of mathematics. It happened in the 1930s and, at that time, platonism was not opposed to nominalism but to intuitionism, and more particularly to questions about totalities. Paul Bernays, who first spoke of platonism, wrote of two opposed *tendencies* in thinking about totalities. The two words that I have italicized, 'attitudes' and 'tendencies', both taken from the authors whom I cite, are in my opinion the right words for the so-called metaphysical issues. I hope they encourage one party to respect the aspirations of the other party in these debates.

In Chapter 7 I shall go backwards in time, to two earlier mathematicians who disagreed profoundly about the nature of their inquiries: Dedekind and Kronecker. Their ideas are intrinsically interesting, but also, at one remove, it is striking how they cut the metaphysical pie rather differently from Gowers and Connes. This may teach a lesson about mathematical metaphysics.

Only at the end of Chapter 7 will I turn to recent philosophy. It is a fascinating field of research, with a gamut of nominalisms endeavouring to circumvent platonism. We will notice a contrast. The Platonism and the anti-Platonism of the present chapter matter to how some mathematicians conceive of what they are doing. They are philosophies of the mathematical life. The platonism versus intuitionism to be introduced in Chapter 7A makes a difference to what mathematicians do. But the platonism/ nominalism debates of contemporary philosophy, discussed in 7B, seem to me to be somewhat to the side. They make little difference to anything outside their own discourse.

A ALAIN CONNES, PLATONIST

8 Off-duty and off-the-cuff

My twenty-first-century pair consists of Platonist Alain Connes (b. 1947) and the thoroughly non-Platonist Timothy Gowers (b. 1963). I will top them up with the late André Lichnerowicz, whom we have encountered in §4.14 for his unusual suggestion about the origin of proof and mathematical objects – the Eleatics. He will be our spokesman for structuralism, in the sense of Bourbaki (§2.12), and not that of contemporary analytic philosophy. It is 'radically non-ontological' (CL&S 2001: 25).

In this chapter I shall speak of capital *P* Platonism, not little *p* platonism, because what the three mathematicians mean by Platonism is reminiscent of the historical Plato as commonly understood. More importantly, it has little to do with the often semantic concerns of contemporary philosophical platonists and anti-platonists. Finally, it allows me to quote, because Gowers explicitly writes with a capital *P*, as do the translators of Connes and Lichnerowicz.

In our days, many mathematicians – not just those who engage with the Foundations of Mathematics or Proof Theory – continue to have strong philosophical opinions about what they are doing, but not many commit themselves in public, let alone in print. We have quoted a few outspoken observations, including William Thurston, 'On proof and progress' (§2.15–2.18, and especially §2.20) and Doron Zeilberger on mathematics being a 'fluke of history' (§2.22). But seldom do we find explicit and to some extent specific published avowals or rejections of Platonism in mathematics. One such is Reuben Hersh, in his lively *What is Mathematics, Really?* (1997). Hersh, to mix metaphors, is the mother of all gadflies, who can't abide the complacent Platonism that he attributes to most of his colleagues. (I know of no statistical evidence that this attribution is correct.) William Tait (2001), already mentioned in §2 above, runs through a series of propositions with which Hersh characterizes common 'Platonist' beliefs, and deflates them from 'super-realism' to what he takes to be a very ordinary 'realism'. I prefer to look at the forthright 'off-duty' but less polemical remarks of three other mathematicians.

We have already encountered Connes in §§ 3.7–9, where he used the Monster to illustrate how a mathematical fact can be just waiting for us, totally unexpected. And, more briefly, Timothy Gowers, in §2.18, where he went out on a limb and suggested that computers might replace mathematicians within a century. And in §3.17, where he was willing to accept the label 'naturalism'. I shall deliberately use rather impromptu statements by both men, because often what is said relatively off-the-cuff, but approved for publication, is more forthright than nuanced. Both men, as already noticed, have the cachet of being Fields medallists.

Connes was asked what he felt about 'off-duty' work: for example, his contribution to Changeux and Connes (1989). He replied that 'it is a way to follow up more general reflection on mathematical activity. I think it is useful work, although secondary' (Connes 2004). Mathematical activity, *l'activité mathématique*. Gowers is quoted in §3.17 as saying that a proper philosophical account of mathematical practice should be grounded in the actual practice of mathematicians. The two men completely agree on *that*. When push comes to shove, they regard the various -isms such as Platonism and structuralism and naturalism and formalism less as doctrines than as *attitudes*. (For Connes' use of the word 'attitude', see §9 below, and for Gowers', §17.) After those points, agreement seems to come to an end.

9   Connes' archaic mathematical reality

I shall also be using an interview Alain Connes (2000) gave to *La Recherche*, a general-interest science periodical that one might file alongside *Scientific*

American rather than the weekly *New Scientist*. In debate with Jean-Pierre Changeux (§3.13), he exemplified the sense that mathematics is just 'out there' to be explored. But Connes is not an indiscriminate 'Platonist'. (Nor was Plato.) He agrees that most of the tools devised by mathematicians are inventions, not discoveries. His label for such tools is 'projective'. They are used to investigate what he calls *archaic mathematical reality* (Changeux and Connes 1995: 192). That's what's thought of as 'out there'.

There is a translation problem here: *archaïque*. The translator of Changeux and Connes (1995) uses 'archaic', while the translator of another debate, *Triangle of Thoughts* (CL&S 2001) uses 'primordial'. This debate was the subject of the magazine interview with Connes that I am using, and so I shall often refer back to it. When I translate from Connes (2000) I shall, like the translators of Changeux and Connes (1995), use 'archaic' – partly *because* it is the more unfamiliar word in English.

Here I may say a few more words about André Lichnerowicz, the *L* of CL&S. He was a senior colleague of Connes at the Collège de France, a distinguished differential geometer and mathematical physicist. He was much influenced by the Bourbaki view of mathematics as the study of structure. A report to the French Ministry of Education, prepared under his direction, strongly influenced, for a while, the teaching of mathematics. He urged that elementary instruction begin with set theory and logic, and should emphasize that maths is about structures. This point of view, which became known in the US as the 'new math', was urged upon American elementary schools in the 1960s. There it gained added impetus from the desire to reform mathematics teaching in order to catch up to the Russians in the aftershock of Sputnik (1957). (See Mashaal 2006, ch. 10 for Bourbaki, Lichnerowicz, and the new math, both French and American.)

The third debater, Marcel-Paul Schützenberger (1920–96), was a polymath mathematical biologist and broader intellectual. Like Karl Popper, but for reasons connected with probability theory, he was highly critical of much current Darwinism, which in turn created some notoriety around his name. He cheerfully describes himself as 'Pythagorean' (CL&S 2001: 27).

To return to 'archaic', it is a relatively new word in both French and English (citation of 1832 for English, 1776 for French). In French it reminds us of the prefix 'archaeo-', as in archaeology – I doubt that it does so for most English readers. The word appears to have entered French with 'archaism', referring to the use in contemporary speech or writing of an old or almost obsolete usage. English followed suit. In the *OED*, the entry for 'archaic' barely fills half my computer screen, while that for 'archaïque' in the best French equivalent, the *Trésor de la langue française*, fills at least three consecutive screens. The two words are not exactly *faux amis* ('false friends'), but the French word has vastly more uses and connotations, and is much more conscious of the original Greek, the ancient spring, beginning, or origin of things.

The word has other uses too, of course. 'Archaic', meaning pre-classical (primarily Greek pre-classical, e.g. Homeric), comes from German art historians who named the epoch *Archaik*. The German *archaische/Archaik* came, like the English word, from the French, and had primarily a philological sense. In conversation about 1980, Michel Foucault assured me that the 'arch' in his word 'archive' in *The Archaeology of Knowledge* was intended to recall the meaning of a source, and origin – and not just a collection of old documents. That meaning is not heard in the English word 'archive'.

Inevitably people have asked Connes what he meant. He expatiated in Connes (2000). I repeat, this *is* an interview, although the interview is more of an art form in French than in English. The interviewee can check and approve every word used by the interviewer in print.

> Although not all mathematicians recognize it, there exists 'an archaic mathematical reality'. Like the external world, this is *a priori* non-organized, but resists exploration and reveals a coherence. Non-material, it is located outside of space and time.

What does 'a priori non-organized' mean? It means that when we encounter it, this primordial reality has no structure. The organization that we attribute to it is a consequence of our conceptualization. Connes thinks the same of the material world around us; what we encounter is unstructured and we organize it with concepts that we project on to it. But we cannot organize any old way. The material world and the archaic mathematical reality are highly resistant to conceptualization; they have their own coherence that we have to respond to. This is my way of reading Connes, and may well be wrong. There is a lot of old philosophical baggage lurking in the attic, and I may have picked up the wrong valise.

At any rate, when we start to examine mathematical reality, we are strongly constrained by what we encounter. 'My attitude and that of other mathematicians consists in saying that there exists a mathematical reality that precedes the elaboration of concepts' (Connes 2000).

> I make an essential distinction between the object of study, for example the series of prime numbers, and concepts that the human mind elaborates in order to understand that series. Archaic mathematical reality is the object of the study. Just like the external reality perceived by the senses, it is *a priori* unorganized. It is sharply distinguished from the concepts of the human

mind elaborated in order to understand it, to see what it has organized. (Connes 2000)

I do not know if Connes would count groups as archaic, or part of the projective structure that we impose on the original reality. What he wrote about the Friendly Giant (or Monster) suggests that how we investigate it, and the concepts we use to describe it, are projected on to reality, but that the gigantic group is in some way part of the coherence of the underlying reality that we encounter. Kant might have called it noumenal. We should begin not with these rather sophisticated questions, but with the very first query. The series of numbers is archaic, primordial, given. 'What I call "primordial [*archaïque*] mathematical reality" is essentially limited to arithmetic truths' (CL&S 2001: 29).

That does not mean the set of provable truths, but a 'Platonic' totality of facts about numbers. Any standard-issue platonist will conclude that if the series of whole numbers is out there, then the *set* of positive integers is also out there in archaic reality. I suggest not. As I understand Connes, sets, and the very concept of a set, are invented by us. We project the set idea on to the series of whole numbers. That does not mean that the set of positive integers does not 'exist', only that it is not archaic or primordial. This distinction is one of the reasons that I find Connes an 'interesting' Platonist.

A quite different attitude to the numbers is that of Kronecker, discussed briefly in §§7.8–10. Like Connes, he thought of the series of whole numbers as given, but, like many of his German contemporaries, he thought of most mathematical concepts and objects as created by us. Yet he appears to have limited the domain of truths about numbers to what can be proven using only those numbers. Kronecker thought that we should not use any of what Connes calls projections, not even the set of positive integers. Thus, starting with a premise very much like that of the Platonist Connes, he is the ancestor instead of constructive mathematics, and, to a certain extent, of intuitionism. Connes' doctrine of archaic reality is curiously akin to Kronecker in one respect, but it is not in the least puritanical, and is thereby akin to Kronecker's rival, Dedekind, in another respect. Dedekind favoured the use of any of our mathematical creations (read projections).

10 Aside on incompleteness and platonism

Could we use a word other than *archaïque*, asks the interviewer. 'Sure, one can speak for example of primitive mathematical reality. The essential is to understand that it precedes the exploration that one is going to conduct – a little like external reality. It is the Platonist (*platonicienne*) position, renewed by Gödel's theorem.'

Connes believes that Gödel's first incompleteness theorem supports his Platonism, on the grounds that there are discoverable truths that lie beyond any one consistent axiomatization adequate for the expression of recursive arithmetic. Thus the theorem is not to be understood as an argument *for* Platonism, but as a fantastic new insight into the nature of mathematical reality, 'out there'. It is so out there that it cannot be captured by any recursive system of axioms.

Gödel was an avowed platonist, but I know of no occasion when he invoked his incompleteness results to vindicate his idea. I have already suggested in §1.20 that he was less a platonist than a leibnizian, believing that once we had adequate ideas (a far deeper understanding of sets than is at present available), we would, for example, be able to settle the truth of the axiom of choice and the continuum hypothesis. But I do not profess Gödel scholarship.

Contemporary platonists in analytic philosophy do not in general make use of incompleteness results to argue for their doctrines. Connes is a Platonist. He thinks there is a totality of arithmetical truths, simply given with the number series itself. Thanks to Gödel we know that totality cannot be characterized by any recursive axiom system adequate to express its own syntax. This is not an argument for Platonism. It is an enrichment of Platonism with a new depth of understanding.

As an attitude to reality and to incompleteness, this seems to me to be impeccable. But to avoid misunderstanding, I have to repeat that as an argument for the existence of an archaic arithmetical reality, with all its truths intact, it begs the question. If you don't think of that reality as already given, then for any consistent and adequate axiom system, Gödel exhibits one sentence which we have grounds to call true, but not provable in that system. This does not show that there is an actual totality of all arithmetical truths.

11 Two *attitudes*, structuralist and Platonist

The interviewer observes to Connes that, in the course of their trialogue, André Lichnerowicz asserts: 'Mathematicians have therefore learned that "the being of objects" [*l'être des choses*] upon which they exercise their reasoning is of no importance.' He also says, 'when we do mathematics, Being with a capital "B" [*l'Être avec un grand Ê*] is disregarded' (CL&S 2001: 24). (For the context of this remark, see the final passage of Chapter 7 below.) This comes directly after the quotation given already in §4.12. We also quoted him there as saying that after Galois and the first glimpses of group theory, 'one of the obstacles to the development of Greek mathematics – the existence of mathematical "objects" – was thereby overcome'. Mathematical objects, once thought to exist, were a positive hindrance to development!

Connes retorts that there are two attitudes to questions about mathematical reality. One is that of Lichnerowicz, 'formalist or structuralist':

> It consists in treating mathematics as a system of logical deductions obtained at the interior of a language, starting from axioms. This position leads in a certain way to denying the ontological character of mathematical reality. The question of the meaning (*signification*) of mathematical objects is evacuated. My attitude and that of other mathematicians consists in saying that there exists a mathematical reality that precedes the elaboration of concepts. (Connes 2000)

That characterization of the difference between the two philosophical attitudes seems exactly right.

12   What numbers could not be

That is the title of a famous paper by Paul Benacerraf (1965).* He observed that current distinct standard set-theoretic definitions of numbers are non-equivalent, and inferred that none of them can be the correct, true, definition of number. The point is also made, but chiefly in the context of Frege's account of numbers as objects, in Parsons (1964: 184–7).

Thus far, Connes would completely agree. If he were to address the matter, he would admire Benacerraf's title. Benacerraf showed that numbers could not be, for example, any of those set-theoretically defined entities. But Connes would go where Benacerraf would not, saying that the integers are prior to any of those set-theoretic entities, and more than any of them. The multiplicity of possible definitions of numbers shows that whole numbers are more than the structures we find in them. Those structures are, in Connes' terminology, 'projective'; they are what we develop as tools to explore the sequence of integers. Many readers of Benacerraf have drawn the opposite conclusion: that all the integers are is structure.

According to Mark Wilson (1992), the founding fathers such as Frege never imagined that they had produced uniquely right definitions of number, although 'Frege must have thought that his own choice was more natural than any alternative' (Parsons 1964: 184). Indeed Quine (1960: 263) suggests 'how one might still argue for the intuitiveness of Frege's version'.

The 'uniqueness problem' did not, I believe, seriously cross Frege's mind. He and others wanted to give explanatory analyses of important concepts, and did not suffer from any delusions that may have afflicted some later analytic philosophers, that analyses aimed at a unique revelation of firm and pre-given meanings. Quine writes with his usual exemplary clarity about this.

> One uses Frege's version or von Neumann's or yet another, such as Zermelo's, opportunistically to suit the job in hand, if the job is one that calls for providing a version of number at all ... That all are adequate as explications of natural numbers means that the natural numbers, in any distinctive sense, do not need to be reckoned into our universe in addition. Each of the three progressions or any other will do the work of natural numbers, and each happens to be geared also to further jobs to which the others are not. (Quine 1960: 263)

Quine's conclusion is that we need not include the natural numbers in our ontology: 'explication is elimination'.

Frege was deeply committed to the idea that numbers are 'objects', and Benacerraf's primary target, in 1965, was Frege, at that time revered by most analytical philosophers of mathematics. And Benacerraf argued (1965: 70): 'Therefore, numbers are not objects at all, because in giving the properties (that is, necessary and sufficient) of numbers you merely characterize an *abstract structure* – and the distinction lies in the fact that the "elements" of the structure have no properties other than those relative to the other "elements" of the same structure.' For a defence of Frege's conception of 'objects' against Benacerraf (1965 and 1973), see Wright (1983).

The Platonist may be unmoved. The series of numbers is primitive. Certainly, in giving their properties you do no more than characterize abstract structures, the very things we use to investigate the numbers. But that is no argument that the numbers are not prior to the structures that we project on to them in order to understand them better.

Unsurprisingly, structuralist mathematicians like Lichnerowicz agree completely with Benacerraf. They do not care what the numbers – or any other mathematical 'objects' (their shudder quotes) – *are*. In recent times, analytic philosophers who profess another type of structuralism (§7.11) conclude that numbers can be no more than whatever structural properties are shared by viable definitions of numbers. Numbers are 'places' in structures (Shapiro 1997). Oddly, they seem still to be locked into the question of saying what the numbers truly *are*.

A Platonist might be tempted to repeat Bishop Butler's slogan 'Every thing is what it is, and not another thing' (Butler 1827: preface). That fits well with Platonism about numbers, but is out of context. G. E. Moore, using it as the epigraph for *Principia Ethica* in 1903, knew his Butler, and knew that it was part of Butler's elegant and powerful statement of what, after Moore, is called the naturalistic fallacy in moral philosophy. Even so, the Platonist will say that the number 17 is what it is, and not another thing. (Certainly not a set or a class or a place in a structure.)

And Quine? A Platonist cries out in amazement: 'Explication is elimination? What on earth is the argument for that?' Within Quine's philosophy of regimentation it makes sense. A Platonist, of course, finds nothing attractive in Quinean regimentation.

The 'elimination' of the natural numbers will return to haunt us in Chapter 7B, when we turn to contemporary platonism/nominalism debates in the philosophy of mathematics.

13   Pythagorean Connes

Connes is firm in his statement that there is a mathematical reality much like external physical reality, but outside of time and space. He does not end with mere analogy:

> I believe that mathematical reality still has many great surprises in store for us. I am ready to say that someday we will recognize material reality is in fact situated inside mathematical reality ... I've already suggested that at the end of my book with Changeux. I believe that one of the criteria of a true understanding of the external physical world, is our capacity to understand it within the mathematical world. We are still far from that. But there are abundant signs. For example, the periodic table of elements. It was deduced by Mendeleev starting with experimental results in chemistry, but when one understands that in fact it is a consequence of extremely simple mathematical facts, it is truly impressive. (Connes 2000)

The interviewer understands Plato better than most, for he interjects, 'That takes you back to the roots of the deepest Platonism!' Connes agrees. Both know that Plato was a Pythagorean. Few platonists among analytic philosophers will confess to being pythagorean, but the sentiment is more widespread among mathematicians than many philosophers have noticed. It is also to be remembered that Connes' memorable work on non-commutative algebra is motivated by quantum mechanics, as he himself asserted in another interview (Connes 2004). We observed in §5.18 that one of his contributions to a rigorous understanding of renormalization is published in *Communications in Mathematical Physics*. His is physics with a pythagorean tinge.

A fan of Alain Connes, cynical about naturalism in mathematics, might observe how flabby, or how ambiguous, the label may be. All self-avowed naturalists are anti-Platonist, anti-platonist, or whatever, for sure. But a disciple of Connes' philosophical views might suggest that *he* is the one who is investigating Nature, properly understood. Most so-called naturalists are mere empiricists, with an impoverished conception of Nature, or so the Platonist might urge. Pythagoreanism is the deeper naturalism.

I do not share Connes' convictions, but my personal opinions are of no importance. We are concerned with a perennial issue. Connes' attitude will persist, I hazard, through generations yet unborn. It is an instructive response to the very experience of many mathematicians. But it is not the only response, as Lichnerowicz attests. Here is another alternative, the more detailed expression of an attitude akin to, but by no means identical to, that of Lichnerowicz.

B   TIMOTHY GOWERS, ANTI-PLATONIST

14   A very public mathematician

Tim Gowers was mentioned briefly in §2.18 for his suspicion that computers will replace mathematicians in a century, and for his internet collaborations, notably the Polymath Project. Aside from his important contributions to combinatorial mathematics, he wrote *Mathematics: A Very Short Introduction* (2002), a model of clarity and brevity. He is the editor-in-chief of *The Princeton Companion to Mathematics* (2008), a remarkable compendium of facts, ideas, and opinions. He is a leading advocate of open access publishing, and organized a boycott of Elsevier journals. He argues that not only are Elsevier's mathematical publications incredibly expensive, but also, he complains, the firm obliges libraries to buy *all* their journals in a field if they buy any (a practice that he says is called 'bundling').

In short, Gowers is very much a public mathematician, very willing to make public statements, and even to engage with philosophy. He was knighted in the Queen's Birthday Honours List, June 2012, 'for services to mathematics'. All of which may make him a bit suspect to some other mathematicians.

Analogous to Connes' interview, I shall use a talk Gowers gave in 2002 to inaugurate a new philosophy of mathematics discussion club at Cambridge University. I do not usually cite URLs, since they tend to impermanence, but in the case of Gowers, advocate of open access, it would be churlish not to provide the address (still accessible August 2012) of 'Does mathematics

need a philosophy?': www.dpmms.cam.ac.uk/~wtg10/philosophy.html.
Page references are to the printed version, Gowers (2006).

Gowers is in one way a completely atypical mathematician: he takes
Wittgenstein's *Philosophical Investigations* (*PI*) very seriously when thinking
about maths. He repeated this in his talk: 'I should confess that I am a fan
of the later Wittgenstein, and I broadly agree with his statement that "the
meaning of a word is its use in the language" (*PI* §43 – actually Wittgenstein
qualifies it by saying that it is true "for a large class of cases").' We shall
return to that maxim in §23 below.

15   Does mathematics need a philosophy? No

Yes and no, Gowers replies. His primary answer is that mathematics does
not need a philosophy. He means that philosophy makes almost no difference to what mathematicians do. Even if tomorrow a philosopher injected
new ideas that completely changed present philosophical thinking about
mathematics, it would make no difference at all to mathematical activity.
He implicitly agrees to a modification of this sweeping claim. I would put it
this way. Logic, Foundations of Mathematics and Proof Theory have long
been branches of mathematics, many of whose practitioners have strong
philosophical motivations, and whose research might well be changed by a
philosophical epiphany. But that would be peripheral to most mathematical
practices. So amended, his claim is almost certainly true today. Philosophy,
as we will confirm in §7.7, mattered more to some mathematicians, in, say,
1884 than 2014.

Gowers cites Wittgenstein (very much out of context!) to convey his
sentiment: 'A wheel that can be turned though nothing else moves with it,
is not part of the mechanism' (Wittgenstein 2001: §271). The quotation is
truncated: the original remark was in Wittgenstein's try-out mode. It began,
'Here I should like to say'. 'Here' was in connection with what is called 'the
private language argument'. However, on the topic of the present chapter,
Platonism, Gowers does seem to me to hit the nail on the head for most
contemporary mathematicians:

> the point remains that if A is a mathematician who believes that mathemat-
> ical objects exist in a Platonic sense, his outward behaviour will be no
> different from that of his colleague B who believes that they are fictitious
> entities, and hers in turn will be just like that of C who believes that the very
> question of whether they exist is meaningless. (Gowers 2006: 198)

By 'outward behaviour' Gowers means, of course, behaviour in mathematical exploration and presentation of results, not what I called 'off-duty'

work. Thus Connes and Lichnerowicz may have disagreed about mathematical 'reality' but, Gowers would insist, that made little difference to how they did their work. Yes, there are quite distinct styles of mathematical practice (Mancosu 2009b), but Gowers believes that the style in which one works is not influenced by philosophical attitudes.

I am not so sure about that, but I have no strong evidence to the contrary. That is a serious question for the sociology of mathematical practice. Do avowedly platonistic mathematicians *today* do mathematics differently from those who disavow platonistic roots?

I emphasize 'today' because, in the great days of disputes about 'foundations', intuitionists really did try to do things differently from the platonists. As we shall see in Chapter 7A, the very word 'platonism' was coined to mark that difference. But suppose those old debates are relegated to history. There are, nevertheless, really contemporary heated disputes between mathematicians that appear to turn on philosophical thoughts. We need go no further than Connes himself. He is wholly sceptical about non-standard analysis. He has criticized it often, on grounds connected to the Platonism described earlier in this chapter – for a sampling, see CL&S (2001: 15–21). Non-standard 'infinitesimals', he claims, simply are not mathematical objects. His position has been rather polemically attacked by Kanovei, Katz, and Mormann (2013). They give many more references to Connes' argument, give their version of its history, and blame it all on his Platonism. This really is a matter of how mathematicians are willing to practise mathematics, and looks like a counter-example to a variant on Gowers' thesis that philosophy makes no difference to how mathematicians do mathematics.

16   On becoming an anti-Platonist

Gowers tells us about his own 'conversion from an unthinking childhood Platonism' (2006: 193). It was when he learnt that the continuum hypothesis is independent of the other axioms of set theory. If as apparently concrete a statement as that can be neither proved nor disproved, then what grounds can there be for saying that it is true or that it is false? (Interestingly, his talk included remarks about the axiom of choice, which one might have thought was in the same boat.)

Well, that is one way to react to the phenomenon: the no-fact-of-the-matter idea. After Gödel showed that the continuum hypothesis was consistent with standard axioms of set theory, Paul Cohen (1934–2007) showed that its negation was also consistent. For this he won the Fields Medal in

1966 – the only Fields Medal ever in mathematical logic. But then he did solve Hilbert's problem no. 1, albeit in a negative way. In the final pages of his own book on the subject, Cohen (1966: 151) writes that 'A point of view which the author feels may eventually come to be accepted is that the CH is obviously false' (original underlining: the book was published from a typescript). He meant, in the simple social sense, that in due course the hypothesis would be unanimously disbelieved, not that it would be disproved from deeper axioms.

Gödel enormously admired Cohen's proof and thought it the best possible. But we all know that, unlike Cohen, Gödel thought the continuum hypothesis was so clear and intelligible that there had to be underlying facts that validated it (or not).

To be moved, and thereby converted, is one thing. Gowers' way is not the only way to be moved. Cohen and Gödel had exactly the opposite take on the question. That is how it is with the Platonist and anti-Platonist attitudes: you can go either way with a truly fundamental fact.

17 Does mathematics need a philosophy? Yes

Gowers names various -isms early in his paper, and variously calls them 'schools of thought', 'positions' and 'attitudes' (2006: 183f). At the end of the paper he advances 'a rather cheeky thesis: that modern mathematicians are formalists, and that it is good that they are' (2006: 199). Formalism has meant quite a few things from the early days of the twentieth century. Gowers is not referring to any precise doctrine that would pass a philosopher's scrutiny, but rather to an attitude to mathematical activity.

He explains that 'Formalism is more or less the antithesis of Platonism'. He starts with what he calls a caricature of the process of mathematical research, and concludes:

> At the end of this process, what we know is *not* that the theorem is 'true' or that some actually existing mathematical objects have a property of which we were previously unaware, but merely that a certain statement can be obtained from certain other statements by means of certain processes of manipulation. (2006: 183, my emphasis)

What is his evidence that modern mathematicians are formalists? (Reuben Hersh thinks just the opposite! A certain lack of data here, not so easily remedied as one might think, for what is the question?) True to his naturalist attitude, Gowers turns to what mathematicians say to each other when they discuss unsolved problems: 'what they are doing is not so much trying to uncover the truth as trying to find proofs'. They devise lemmas and debate them. 'The probable truth and apparent relevance of the lemma are basic minimal requirements, but what matters more is whether it forms part of a realistic-looking research strategy, and what that means is that one should be able to imagine, however dimly, an *argument* that involves it' (2006: 199). (The Platonist agrees!)

Gowers suggests that 'most successful mathematicians are very much aware of this principle', and that 'it is a good idea to articulate it . . . And it is a principle that sits more naturally with formalism than with Platonism' (2006: 199). He also argues it is good pedagogy. What matters in teaching definitions 'is the basic properties enjoyed by the objects being defined, and learning to use these fluently and easily means learning appropriate replacement rules rather than grasping the essence of the concept. If you take this attitude to the kind of basic undergraduate mathematics I am teaching this term, you find that many proofs write themselves.' Here we notice the Wittgensteinian background. What the pupil learns is the use – the fluent use – of the terms in the mathematical language, and one should resist the temptation to say anything more about their 'meaning' than that.

In connection with pedagogy, a recently published essay (Gowers 2012) contrasts two ways of introducing the concept of a group, one by abstract definition, and the other using a vivid story to explain what the point of thinking about groups is. We are left to see, partly by comparison with ordinary story-telling, which is the most effective. One might add, that is true for most learners, but not for all. I can think of one major group theorist mentioned in another chapter who probably preferred to be told about groups right off, in a literal and abstract way.

Finally, Connes called Lichnerowicz 'formalist or structuralist' (§11 above). I think we can say that he was pretty much of a formalist in Gowers' sense too. The two anti-Platonists come from different traditions, and write in different idioms, but their attitudes are quite similar. In one respect they might well differ: teaching.

> One of the standard tricks that we do as mathematicians is 'reduce' one concept to another – showing, for example . . . that positive integers can be 'built out of sets'. People sometimes use extravagant language to describe such constructions, sounding as though what they are claiming is that positive integers 'really are' special kinds of sets. Such a claim is, of course, ridiculous, and probably almost nobody, when pressed, would say that they actually believed it. (Gowers 2006: 189)

In another connection, he rejects the idea that 'an ordered pair is "really" a funny kind of set. (*That* view is obviously wrong, since there are many different set-theoretic constructions that do the job equally well)' (2006: 192). (Compare Benacerraf 1965.) He adds in parenthesis that if you teach undergraduates that ordered pairs are *really* sets in a certain set theory, 'you will confuse them unnecessarily' (2006: 199). This seems very much at odds with the 'new math' pioneered by Lichnerowicz, in which one starts children on sets at the very beginning.

18 Ontological commitment

'One view, which I do not share, is that at least some ontological commitment is implicit in mathematical language' (Gowers 2006: 189). The famous phrase is Quine's. It is essential to contemporary platonism/nominalism debates to be discussed in Chapter 7B. The phrase itself is discussed in §§7.18–20. Here it may suffice to say that Gowers is an anti-Platonist, but that does not make him into a nominalist in the sense of those debates. The contemporary nominalist says numbers do not exist. Gowers does *not* deny that numbers exist. He thinks it would make no sense to assert that numbers do exist – or that they do not exist.

In his classic 'On what there is', Quine said that the 'mediaeval points of view' that historians call *realism, conceptualism*, and *nominalism* 'reappear in twentieth-century surveys of the philosophy of mathematics under the new names *logicism, intuitionism*, and *formalism*' (Quine 1953: 14). So he thought of formalism as the continuation of the nominalist instinct. This all fits together, given Gowers' professed formalism. Lichnerowicz avowed finding Ockham more interesting than Abelard. His Ockham thought that 'the realists and nominalists are fighting over nothing' (CL&S 2001: 37). His Ockham sounds a bit like my resurrected and reconstructed nominalist (end of §7 above). As I understand Lichnerowicz, we are not to assert or to deny that numbers or other abstract 'objects' exist. That is to fight over nothing. Look instead, Gowers would advise, at how the names for these things are used.

Although I would prefer silence, one ought to say something to deflate the following banal argument: Are there infinitely many primes? – Yes! And prime numbers are numbers? – Well, uh, what does that mean? Of course prime numbers are numbers, but what's being said here? – Well, if prime numbers are numbers and there are infinitely many primes, numbers exist!

Gowers avails himself of a distinction made by Rudolf Carnap (1950). There are internal and external questions. The statement about primes is an assertion *internal* to mathematical discourse. Likewise if I say that there are at least forty-seven perfect numbers (§5.17, *App 5*), of which the fifth is

33550336. The dialogue about primes ends with an assertion *external* to that discourse, the statement that numbers exist. As Gowers puts it, 'the answers you give in the internal sense do not commit you to any particular external and philosophical position'. There is ever so much more to say, books and books of philosophical analysis, so Gowers was prudent to stop there. He is opting out of the ontological commitment debates.

I was reminded of the absurdity of external discourse when in Cape Town I dialled a wrong number. A very firm recording announced: 'The number that you have called does not exist.' Now *that's* non-existence of a number. Internal to the discourse of South African telephone numbers. Suppose I had been talking in the good old days to a person, not to a machine. I look at my address book and read out 650 3316: 'You're telling me that number does not exist?' Operator replies, 'Yes *that* number exists, try dialling it again.' Operator is not committed to any philosophical position.

But that is not quite the end of the matter. Pythagoreans have an interesting take. The only numbers that are real are those that are deeply embedded in physics. The third figure on the *Triangle of Thoughts*, Paul Schützenberger, held that, 'in my opinion, outside of physics, there is no such thing as an integer; consequently, we are engaged in empty mathematics' (CL&S 2001: 90). He calls this a Pythagorean thesis. What impresses him is that, for example, in crystallography there are certain intrinsic whole number relationships. That's where integers live. And if there is an integer that has no place in the world, then it does not exist. So some integers may not exist – in its context, an 'internal' context, a statement on a par with the South African recorded voice telling me that the number I have dialled does not exist.

19   Truth

Anyone present at Gowers' talk would have noticed that he quite often used (and did not just mention) the adjective 'true' and, indeed, 'fact of the matter'. Isn't he talking like a Platonist? 'Another conjecture that seems almost certainly true is the twin primes conjecture – that there are infinitely many primes p for which $p+2$ is also prime' (2006: 196). What is it to 'seem almost certainly true'? The curt answer is it looks like it is provable. Recall some words of his quoted in §17 above: mathematicians discussing conjectures are 'not so much trying to uncover the truth as trying to find proofs'. What matters 'is whether it forms part of a realistic-looking research strategy, and what that means is that one should be able to imagine, however dimly, an *argument* that involves it'. At any rate, as philosophers have endlessly argued, the use of the adjective 'true' does not commit the speaker to a correspondence theory of truth, which is what you would need to saddle Gowers with Platonism.

Gowers characteristically looks at the particular case. Why does he feel that the twin primes conjecture is true? He mentions heuristic considerations. The primes appear to be 'distributed randomly'. (If twin primes ran out, they would not be.) A 'sensible-looking probabilistic model for the primes not only suggests that the twin primes conjecture is true but also agrees with our observations about how often they occur'. This is how the conversation goes on. I believe that every time Gowers uses words like 'true' or 'fact of the matter', one can rephrase what is said in a more cumbersome way.

20   Observable and abstract numbers

Gowers continues with a question about quantifiers. The primes conjecture is of the form, 'for every so and so, there is a such and such'. He is pretty sure there is a 'fact of the matter', that there are at least n twin primes, for any n you might choose to specify. 'Any' is a little different from 'every'. He is more cautious about saying that for every n there are at least n twin primes. Perhaps this is another case where he might agree that mathematics needs help from mathematical philosophy, because $\forall x \exists y$ statements are one of the things that proof theory is all about.

He has his own line on such matters, which, he agrees, is one of two possible attitudes (that word once again). He distinguishes 'observable' from 'abstract' statements about numbers. The 'observable' statements are about numbers small enough to be intelligible, and such that it at least makes sense at present to speak of checking them. But a number of the magnitude of $10^{10^{100}}$ 'isn't really anything more to us than a purely abstract n'. (He probably meant 10 to the 10th power taken to the 100th power.) Hence 'for all n' statements are

> after a certain point, no more concrete than the general statement, the evidence for which consists of a certain manipulation of symbols, as a formalist would contend. So, in a sense the 'real meaning' of the general theorem is that it tells us, in a succinct way, that the small 'observable' instances of the theorem are true, the ones that we might wish to use in applications.

I believe that he had in mind what Manders called math-math applications, not the activities of members of SIAM.

Here is his example. There is a theorem due to Askold Ivanovich Vinogradov (1929–2005, not to be confused with another and more famous

Russian mathematician, Ivan Matveevich Vinogradov, 1891–1988) that states that every sufficiently large odd integer is the sum of three primes. At present we cannot do better with 'sufficiently large' than at least 10^{13000}, and, Gowers tells us, there are only seventy-nine primes known greater than that, so the theorem has almost no observable consequences.

Consider, he says, that staple used by philosophers of mathematics, the Goldbach conjecture that every even number is the sum of two primes. Many mathematicians think that Vingradov's theorem 'basically solves' Goldbach's conjecture. But since we at present have only seventy-nine primes to play with, this hardly helps Gowers at all. Thus, he says, his attitude is different from that of most (pure) mathematicians. His own attitude results from the fact that his speciality of combinatorics proceeds by a lot of experimental mathematics, noticing patterns in the integers, and checking on their general validity. So *he* wants observable numbers one can play with, not abstract numbers which one cannot, at present, manipulate.

Gowers was hardly the first to raise this type of worry. In Chapter 7 we will examine Paul Bernays' coinage of the label 'platonism' in mathematics. In the course of the 1934 lecture in which he did so, he observed that:

> Intuitionism makes no allowance for the possibility that, for very large numbers, the operations required by the recursive method of constructing numbers can cease to have a concrete meaning. From two integers k, l one passes immediately to k^l; this process leads in a few steps to numbers which are far larger than any occurring in experience . . . [He gives the example of 67 raised to the 257th power, all raised to the 729th power.]
>
> Intuitionism, like ordinary mathematics, claims that this number can be represented by an Arabic numeral. Could not one press further the criticism which intuitionism makes of existential assertions and raise the question: What does it mean to claim the existence of an Arabic numeral for the foregoing number, since in practice we are not in a position to obtain it? (Bernays 1983: 265)

Perhaps Gowers' contrast between 'observable' numbers and merely abstract ones is a useful way to carry on that discussion. Gowers' observables today will be far larger than those possible for Bernays, in 1934, because Gowers can use the most powerful computer available today. More important, perhaps, is Bernays' emphasis on the role of, and limits of, our ability to name specific numbers. Our standard systems for writing numerals depend upon an important invention, dating, it seems, from the Sumerians: the use of what we call 'zero' not only as a sign for nothing, but also as a place-marker. In the system of Arabic numerals, that's our sign 'o'. (The Maya had a comparable place-holder.) It is often said that counting 1, 2, 3, ... obviously goes on forever, and therefore we have the concept of infinity. Oh no, counting does *not* obviously go on forever – not without a place-holder (or other invention).

Various names ending in '-ism', such as 'strict finitism', have been proposed to label considerations such as those broached by Bernays and Gowers, but I find it more profitable to reflect on their examples than to formulate yet another doctrine or dogma.

We may add that, in 1959, Bernays (1964), reviewing Wittgenstein's *RFM*, was particularly interested in his remarks about notation and large numbers, observations that leave most philosophers cold, or else lead them to classify Wittgenstein as some sort of 'finitist'. Bernays is closer to the phenomena than most people who read those passages, for he is less interested in '-isms' than in mathematical experience.

21 Gowers versus Connes

I have been treating the two mathematicians as exponents of two different attitudes, and only in imagination pitting them as opponents. There happens to be one explicit confrontation in Gowers' talk. It is not at the heart of their differences, but I should mention it briefly (and out of its context). Connes spoke, perhaps unguardedly, of a 'direct access to the infinite' which aroused Gowers' hackles. Here's Connes:

> this direct access to the infinite which characterizes Euclid's reasoning, or in a more mature form Gödel's, is actually a trait of the living being that contradicts the reductionist model. Roger Penrose has developed an analogous argument. (CL&S 2001: 157, referring to Euclid's proof that there are infinitely many primes; the Penrose argument is in e.g. Penrose 1997)

For philosophers, the argument is reminiscent of John Lucas' essay 'Minds, machines and Gödel' (1961). He argued that minds are not machines on grounds of Gödel and Turing. From the day that was published, most analytic philosophers (including my immature self) were sceptical; see, for example, Benacerraf (1967).

Gowers, always optimistic about computers, reacted as follows to this sort of consideration: 'I have sometimes read that computers cannot do the mathematics we can because they are finite machines, whereas we have a mysterious access to the infinite. Here, for example, is a quotation from the famous mathematician Alain Connes' (Connes' words in CL&S just quoted). Gowers' retort perhaps reflects his preference for looking at how words are used – and what has to be taught in order to use them:

But, just as it is not necessary to tell a computer what an ordered pair is, so we don't have to embed into it some 'model of infinity'. All we have to do is teach it some syntactic rules for handling, with care, the *word* infinity – which is also what we have to do when teaching undergraduates. And, just as we often try to get rid of set-theoretic language when talking about sets, so we avoid talking about infinity when justifying statements that are ostensibly about the infinite. (Gowers 2006: 193)

Gowers is deflationary. Don't get carried away in flights of fancy! Look at how *ordinary* our reasoning is. Look at how the word 'infinity' is *used* in proving that there are infinitely many primes.

Let's examine the maxim to which I'm alluding: don't ask for the meaning, ask for the use. It sounds great, but it is not at all clear how to use it. I find it suggestive rather than rigorous, but none the worse for that. Connes does not to my knowledge use it against what Paul Benacerraf (1973: 664) has called a standard semantical account. (In that paper he always put the word 'standard' in scare quotes.)

22   The 'standard' semantical account

Benacerraf's superb paper, 'Mathematical truth' (1973), is a fundamental benchmark for philosophical platonism/nominalism debates. It is to be read, not summarized. It poses a dilemma, whose two horns are:

(α) A statement such as (2) '*There are at least three perfect numbers greater than 17*', is to be analysed according to standard semantics, or

(β) it is not.

These lead to the dilemma of interest, namely:

(γ) Platonists can tell us what mathematical theorems are about, namely numbers and other 'abstract objects'. Current analytic epistemology holds that we have to have some sort of causal relationship to any object in order to know anything about it. We cannot have a causal relationship with things not in space and time. Therefore platonists can tell us what they are talking about, but not how they can have any knowledge of such 'objects'.

(δ) Non-platonists who emphasize proof can tell us how they know mathematical truths, namely by proof, but cannot tell us what they are talking about.

In the immense literature provoked by Benacerraf's paper, very few authors attend to the first step, the dilemma of α against β. What is a 'standard' semantical account?

It is what I shall call 'denotational', following the practice of a recent paper by two grammarian–philosophers, as they might be called, Christopher Kennedy and Jason Stanley (2009). A more familiar label might be 'referential', but I avoid it for reasons too messy to explain here. Denotational semantics is a way to express word–world relations in an apparently non-committal way, making full use of the repertoire of semantic ascent. A core feature is that it allows our words to hook up with things in the world in a piecemeal way.

It is well suited to a statements such as (1) *'There are at least three large cities older than New York'*. The word 'older' denotes the relation of being older than, and 'New York' denotes a city. Where I use the old-fashioned verb 'to denote' used by Mill and Russell, others would say 'names' 'stands for', 'refers to', or 'designates'. The niceties are irrelevant here. Benacerraf initially avoids all of them, but 'reference' soon enters as the critical idea for which I am using Russell's 'denotation'.

On the surface, (1) – about cities – and (2) – about perfect numbers greater than 17 – look grammatically very similar. Do they have the same 'logicogrammatical form', as Benacerraf calls it? Prima facie a denotational semantics suits (1). Does it suit (2)? If so, then '17' in that statement denotes a number. Of course it does. So what?

That is where the alchemy begins to brew. We want a general label for whatever the term occurring in the 'New York' position denotes. Benacerraf takes the pro-noun 'entity', although others perhaps imprudently speak at once of 'objects'. We pass to the Tarskian truth conditions for sentences of that logicogrammatical form, and ask how we can know those conditions to hold. Only if we know about the entities denoted by, for example, 'New York' or '17'. On a standard view about knowledge, one can have knowledge of an entity only if there is some sort of causal connection between the knower and the entity known. We mortals have causal relationships only with entities in time and space. Hence horn (γ). That's the horn troubling for platonists, content with numbers as abstract objects, but unable to explain how we know about them. Most platonists quarrel with the epistemological premise, arguing that there are many non-causal roads to knowledge.

So how do we get to horn (δ)? By rejecting horn (β), thereby impaling ourselves on horn (α). Benacerraf takes (β) as a viable alternative, and notes that many great thinkers in the past, among whom the most eminent is David Hilbert, have opted for (β). But he settles on a requirement which says, in effect, that our semantics for mathematical language should be uniform with the rest of the language we use. I shall label this requirement 'SU' for semantic uniformity. SU is the requirement that:

the semantical apparatus of mathematics be seen as part and parcel of that of the natural language in which it is done, and thus that whatever *semantical* account we are inclined to give of names or, more generally, of singular terms, predicates, and quantifiers in the mother tongue include those parts of the mother tongue which we classify as mathematese. (Benacerraf 1973: 666)

Benacerraf observed that Hilbert would *not* have agreed to SU. Nowadays, most philosophers who participate in current debates about platonism versus nominalism *do* agree. It is part of the very framework for contemporary platonism/nominalism discourse, to which we return in Chapter 7B.

Wittgenstein asked, 'Is it already mathematical alchemy, that mathematical propositions are regarded as statements about mathematical objects, – and mathematics as the exploration of these objects?' (*RFM*: v, §16b, 274). I have been describing the alchemy by which the dilemma α/β is transmuted into the dilemma γ/δ.

Nobody who takes Wittgenstein seriously is likely to agree to denotational semantics applied to mathematics. This is not only because, after the *Tractatus*, Wittgenstein came to the conclusion that SU expresses a profound misunderstanding of language. It is also because a denotational semantics for maths seems to run counter to the famous maxim that Gowers has cited on several occasions, 'don't ask for the meaning, ask for the use'. Let us see why.

23   The famous maxim

There are waves of enthusiasm for aphorisms. Soon after Wittgenstein's *Philosophical Investigations* was published, one favourite was, 'don't ask for the meaning, ask for the use'. I unkindly filed it with jargon in §1.2, partly because although it sounds brilliant, it is far from clear how to deploy it. Perhaps that is why it is not cited so often nowadays. In contrast, whenever Gowers mentions how much he has learned from the *Investigations*, he cites this maxim. As he pointed out, it is a truncation of what Wittgenstein actually wrote. *PI* §43 reads:

> For a *large* class of cases – though not for all – in which we employ the word 'meaning' it can be defined thus: the meaning of a word is its use in the language.
>
> And the *meaning* of a name is sometimes explained by pointing to its *bearer*.
>
> Man kann für eine *große* Klasse von Fällen der Benützung des Wortes ‚Bedeutung' – wenn auch nicht für *alle* Fälle seiner Benützung – dieses Wort so erklären: Die Bedeutung eines Wortes ist sein Gebrauch in der Sprache. [Both emphases are in the German original, although there is only one in the translation. Perhaps 'explain' or even 'explicate' is preferable to 'define'.]

Und die *Bedeutung* eines Namens erklärt man manchmal dadurch, daß man auf seinen *Träger* zeigt.

I've copied out the German to remind us of Frege's famous paper of 1892, *Über Sinn und Bedeutung*. The word *Bedeutung* is perfectly ordinary German, and Anscombe's translation of *PI* §43 is entirely sound. Frege, however, gave a somewhat idiosyncratic twist to a perfectly ordinary German word. There is some evidence that Russell thought he meant 'denotation', a choice made by Montgomery Furth translating Frege (1967). In Frege (1949), Herbert Feigl chose 'nominatum' when translating *Sinn und Bedeutung*. In Frege (1952), Max Black chose 'reference'. For many years thereafter the famous paper was known as 'On sense and reference'. This was revised in Frege (1980), when it became 'On sense and meaning'. That led Michael Dummett to remark:

> The translation remains the same, save for the rendering of certain of Frege's technical terms. Of these changes, by far the most important is the substitution of 'meaning' for 'reference' as the translation of *Bedeutung*. It is, to my mind, a great pity that this rendering was not adopted in the original edition. Since then, largely through the influence of the Geach/Black translation, *the word 'reference' has become standard, in English language philosophical writing* that discusses or alludes to Frege, and cannot now be dislodged in deference to a change of mind 28 years later, at least without rendering a great deal of commentary unintelligible. (Dummett 1980, my emphasis)

Michael Beaney, editing *The Frege Reader* (1997), threw up his hands and chose to leave the words *Sinn* and *Bedeutung* in the original German!

At any rate, the word 'reference' became standard not only in discussions of Frege, but in analytic philosophy overall. As an immediate example of this usage, Benacerraf himself said he wanted a 'philosophically satisfactory account of truth, reference, meaning and knowledge' (1973: 662), but in a footnote immediately evaded talk of meaning, asserting that 'Reference is what is presumably most closely connected with truth, and it is for *this* reason that I will limit my attention to reference'. In §12 above I mentioned Parsons (1964) as a companion to Benacerraf (1965); the terminology throughout is that of 'reference'.

Wittgenstein happily knew nothing of all that, but he was familiar with Frege's idiosyncratic usage. He used the noun *Bedeutung* often in the *Investigations*, and it is plausible to guess that Frege's usage was sometimes nearby. Start with *PI* §1, which suggests 'a particular picture of the essence of human language':

> the individual words in language name objects – sentences are combinations of such names. – In this picture of language we find the roots of the following

> idea: Every word has a meaning [*Bedeutung*]. This meaning is correlated with the word. It is the object for which the word stands.

Wittgenstein was talking about Augustine and his own *Tractatus*. But might he not have had Frege's early version of denotational semantics also in mind? When 'Jupiter has four moons' is transformed into 'The number of moons of Jupiter is four', with the consequence that in the latter the token of 'four' denotes the object four, might not a reader protest, don't ask for the *Bedeutung*, ask for the use? Gowers (2006, 186) explicitly declined to discuss what he calls 'direct reference', but he could well have invoked the famous maxim.

As his philosophy evolved, Wittgenstein absolutely rebelled against the uniformity-of-semantics premise SU. Hence he had no use for denotational semantics. But his maxim, don't ask for the *Bedeutung*, gives a specific impetus for rejecting it. Indeed it is one of the reasons why although I think that denotational semantics is *one* useful model of at least European languages, I reject it as the uniquely correct way to analyse language. (There is no more a uniquely best way to analyse English than there is a uniquely best analysis of the concept natural number.)

24   Chomsky's doubts

There are very few points on which Wittgenstein and Chomsky can be thought of as in agreement, but Chomsky was also dubious about denotational semantics. That began in the 1950s. His doubts continue. Here is how he recalled them in 2000:

> As for semantics, insofar as we understand language use, the argument for a reference-based semantics (apart from an internalist syntactic version) seems to me to be weak. It is possible that natural language has only syntax and pragmatics; it has a 'semantics' only in the sense of 'the study of how this instrument, whose formal structure and potentialities of expression are the subject of syntactic investigation, is actually put to use in a speech community,' to quote the earliest formulation in generative grammar 40 years ago, influenced by Wittgenstein, Austin, and others [he here recalls his publications from 1955 and 1957]. (Chomsky 2000: 102f)

This view is nowadays very much a minority opinion. Since he was concerned with 'reference-based' or denotational semantics, Chomsky might have mentioned Strawson's 'On referring' (1950) alongside Austin.

25 On referring

In the course of the twentieth century there were three sensational contributions to what we now call the theory of reference. The first was Bertrand Russell's 'On denoting' (1950). Next was P. F. Strawson's 'On referring' (1950). Finally, Saul Kripke's theory of rigid designation (1980). Note the change in wording. Denotation – reference – designation. Russell had taken denotation from John Stuart Mill, who continued an ancient tradition according to which nouns and noun phrases have both denotation and connotation, names only denote. In a few respects, Kripke took the wheel full circle, back to Mill's connotation-free simple denotation = designation.

Russell had taught that names for people with whom we are not acquainted, but know only by description, are concealed definite descriptions, hence to be analysed out, in a way that has become very well known, and which I shall briefly describe in §7.16. For Mill they had simply been names. Kripke extended this idea of common names, or nouns, for substances and much else, in his celebrated theory of natural kinds. In all these turnings of the wheel, Strawson's 'On referring' has tended to be left by the roadside.

Strawson, who was a close colleague of J. L. Austin emphasizing speech acts, argued that Russell's theory of definite descriptions, advanced in 'On denoting', was plain *wrong*. One of the reasons that it was wrong is indicated by Strawson's observation that '"Mentioning" or "referring" is not something that an expression does; it is something that someone can *use* an expression to do. Mentioning, or referring to, is characteristic of a *use* of an expression' (Strawson 1950: 326, my italics). On the next page he explicitly contrasted meaning and referring. I suspect that was one reason Max Black chose 'reference' to contrast with 'sense' in translating Frege. If so, that was exactly to miss Strawson's point, that they are from different categories: that is, cannot apply to the same entity. Expressions *have* meanings, while we *use* some expressions to refer.

Strawson's ideas about presupposition and truth value gaps, presented in his 1950 paper, have been widely assimilated. His point about occasions of use has been well developed in the theory of indexicals and situational semantics. Yet his first lesson, that *words do not refer*, but that *speakers use words to refer*, seems to have been largely forgotten. It is, however, with the memory of Strawson's paper, that I follow Kennedy and Stanley (2009), and call the 'standard' semantical account denotational rather than referential.

I have strayed a good deal from Gower's use of Wittgenstein's maxim about use, as a reminder that there is a respectable tradition alternative to that of the theory of denotation (reference). We shall return to this in Chapter 7B, where we show in §7.20 how Thomas Hofweber's recent work suggests how to avoid ontological commitment to numbers.

The short message is, however, brief. Benacerraf made absolutely clear that his famous dilemma took for granted semantical uniformity and denotational semantics. Without the premise that '17' denotes a number in the way that 'New York' denotes a city, the dilemma does not arise.

CHAPTER 7

Counter-platonisms

1 Two more platonisms – and their opponents

Platonism, as understood by our mathematicians of Chapter 6, holds that there is a mathematical reality, independent of the human mind, which human beings investigate, explore, discover, very much as we gradually learn more and more about the material world in which we live. Alain Connes, self-avowed Platonist, has a deep commitment to the existence of an archaic reality constituted by the series of numbers. The anti-Platonists, Alain Lichnerowicz and Tim Gowers, deny that this 'reality' makes sense. They do not think of themselves as exploring a reality and discovering the truth, but as discovering what can be proven. All three describe their views more as attitudes than as systematic doctrines. These attitudes express their visions of the mathematical life, of what they are doing, as mathematicians.

Now we turn to philosophical doctrines. They run parallel to the mathematicians, but arise from different interests. The platonistic side in the ensuing debates is best understood by what opposes it. I shall distinguish two different types of counter-platonism. The first, introduced in Part A below, is connected with intuitionist and constructivist tendencies in mathematics. It is what led Paul Bernays to introduce the word 'platonism' in connection with mathematics. The second, in Part B, arises against a background of denotational semantics, and is the primary focus of today's platonism/nominalism debates among analytic philosophers.

You might expect this chapter to be directed at the second group, the contemporary debates. They are, after all, at the heart of today's philosophical writing about mathematics. But I pass them by in an all too casual way. Why? Because, as some of the protagonists have stated, these discussions are of very little interest to working mathematicians. No matter what consensus builds (in the unlikely event that there is a progression towards agreement), it is unlikely to make much difference to mathematical activity or to what mathematicians think they are doing. A decent study of the congeries of arguments in circulation demands a different book. Fortunately, any interested reader not already versed in these matters can consult kept-up-to-date surveys of specific programmes readily found in the online *Stanford Encyclopedia of Philosophy*. They are all written by experts in the field, and list the standard literature.

Very well, but why return to the first modern use of the name 'platonism' in connection with mathematics? Because it illustrates another facet of the philosophy of mathematics, once of great importance, but now on a back burner. It also allows us to be even more retrograde, going back to the emergence of that facet, in the persons of two late nineteenth-century mathematicians with intense philosophical views that, as I put it in §6.7, cut the metaphysical pie differently from the way that our contemporaries do. I suggest that this seriously enriches our understanding of what keeps platonistic philosophies of mathematics perennial. Indeed it is part of the full story of why there is philosophy of mathematics *at all*. It confirms the argument of Chapter 3, that one reason is ancient, and has its roots in Plato, and that the problems arise because of the experience of exploration, and proof, or rather the experiences of explorations and proofs, all in the plural, which evolve as the very notions of inquiry and of proof evolve.

I shall, then, be describing two kinds of platonism, one opposed by intuitionist tendencies, and one, more briefly, opposed by a current kind of nominalism. Both are recognizably related to the Platonism discussed by the mathematicians, but invoke other interests and other opponents. The opposing doctrines, the counter-platonisms, are at first sight rather more subtle than the brash anti-Platonism of the mathematicians in Chapter 6.

A   TOTALIZING PLATONISM AS OPPOSED TO INTUITIONISM

2   Paul Bernays (1888–1977)

'Paul Bernays is arguably the greatest philosopher of mathematics in the twentieth century.' That is the first sentence in the 'Project Description' of the Bernays Project at Carnegie-Mellon University. I have no quarrel with that judgement, even though the topics of the present book are far from his contributions. An introduction to his work is given on the site of the Bernays project, at: www.phil.cmu.edu/projects/bernays/.

Bernays was taught in Berlin by, among others, Ernst Cassirer, Edmund Landau, and Max Planck. He went on to Göttingen to study under David Hilbert, Hermann Weyl, Felix Klein, Max Born, and Leonard Nelson. That makes him about the best-educated man of his generation who had mathematical and philosophical interests. He became Hilbert's assistant; he was professor at Göttingen until he was fired in 1933. Happily he inherited Swiss citizenship and could move to Switzerland. His most famous pupil before the war was Gerhard Gentzen. The main alternative to Zermelo–Fraenkel set theory even today is von Neumann–Bernays–Gödel set theory. Tait (2006: 76), writing of the published version of Gödel's letters, notes 'the disproportionate size of the Bernays correspondence': eighty-five letters are included (almost all of them, spanning 234 pages, including the face-to-face originals and translations). Many readers would file him as a logician, but the Bernays project got it right: an exceptional philosopher of mathematics. Finally, one should remember the philosophical milieu in which he lived: phenomenological and neo-Kantian.

Here I am concerned only with Bernays' introduction of the label 'platonism'. Doubtless the word 'platonism', with reference to mathematics, was already in circulation. But Bernays appears to have been the first so to use it in well-preserved print. Bernays (1935a) is the published version of a talk he presented in 1934 in Geneva. It was translated by Charles Parsons (Bernays 1983), who preserved the lower-case *p* of 'platonisme' that is obligatory in French. This is convenient for my convention, but of no significance, because Bernays usually wrote in German, where an upper-case *P* would have been mandated. I may mention that Parsons translates Bernays' *nombres entiers* (1935a: 53) as 'integers'; I have used 'whole numbers' below, which accords, for example, with Kronecker's German usage *ganzen Zahlen* (§7 below).

3   The setting

In 1934 the University of Geneva organized a series of lectures under the general rubric of *La logique mathématique*. It began with a paper by the Belgian philosopher Marcel Barzin (1891–1969). It was titled, 'On the current crisis in mathematics'. Barzin (1935: 5) credits the idea of a 'current' or 'new' crisis to Hermann Weyl's (1921) paper on the *new* foundations crisis (*die neue Grundlagenkrise der Mathematik*). Not, for instance, the old 'crisis' induced by the antinomies.

Weyl's 1918 book on the continuum (Weyl 1987) was a proposal for analysing the continuum in the spirit of what we call, after Brouwer, intuitionism. For the main players in this 'new' crisis (Bernays, Brouwer, Heyting, Hilbert, Kolmogorov, and Weyl), see Mancosu's excellent anthology (1998). Brouwer's intuitionism figures hugely in the published papers from the 1934 lecture series, and Bernays states that Brouwer had spoken to much the same audience in March of that year.

Bernays begins, however, by rejecting the premise of a 'crisis' in general. '*The truth is that the mathematical sciences are growing in complete security and harmony*' (Bernays 1983: 258, my emphasis). He never had reason to change his mind. In 1959 he reviewed Wittgenstein's *Remarks on the Foundations of Mathematics*. Wittgenstein's dismissal of the importance of contradictions has excited much debate among some philosophers. Bernays says only that 'what does not appear from his [Wittgenstein's] account is that contradictions in mathematics are to be found only in quite peripheral extrapolations and nowhere else' (Bernays 1964: 527). Bernays would have been astonished by Voevodsky's thought (§1.22) that today inconsistency remains more than a real possibility.

Note that the paper on platonism was not Bernays' only contribution to the 1934 Geneva lectures. He followed with the modestly entitled, 'On some essential points about metamathematics' (Bernays 1935b). This is a brilliant exposition of the state of the art. I have called the counter-platonism under discussion intuitionistic; in fact, the distinction between 'intuitionistic' and Hilbert's 'finitary' is clearly enunciated for the first time in Bernays' lecture. He proceeded through a very careful account of the results of Jacques Herbrand (1908–31), whose two papers (Herbrand 1967a, b) of 1930 and 1931 were the final contributions to Hilbert's programme before Gödel. He proved the consistency of a major part of arithmetic, and added an appendix explaining why his proof does not conflict with the incompleteness theorems of which he had just been apprised. Herbrand climbed mountains and fell to his death at age twenty-three.

After presenting Herbrand's work (with a mention of the initial results of his own student, Gerhard Gentzen) Bernays proceeds to a lucid exposition of Gödel's results. Dawson (2008: 94) states that doubts about the second incompleteness theorem were not 'stilled' until publication, in 1939, of an impeccable proof in the second volume of the Hilbert and Bernays' classic, *Die Grundlagen der Mathematik*. With hindsight, it is hard to see what doubts might remain in the mind of anyone who attended Bernays' Geneva lectures. Such matters are of fundamental importance to the history of the subject, but in this chapter I am concerned only with something rather trivial – the emergence of the notion of platonism in connection with mathematics.

4 Totalities

Do we have a clear idea of the totality of natural numbers? It seems so simple, 1, 2, 3, and so on – all of them. Bernays suggests we may be using an unjustified analogy. I have just bought a dozen eggs. That's clear. Now I form the idea of a set with exactly those members. Is that a clear idea? Certainly. We can make this move by analogy with the way in which, confronted by a carton of a dozen eggs, we not only speak of each individual egg, identified, say, by its place in the carton, but also form the idea of a carton, set, collection, or group, of twelve eggs. We can certainly reason not only about the eggs but about the set of a dozen eggs.

I have a clear idea of that totality; in fact, I bought the eggs in a carton. And, by analogy, for any number in the sequence of integers, I can form the idea of a totality of all the integers up to that point. Whatever one's worries about abstract as opposed to observable numbers (§6.20), far more problematic is the belief that we have a clear idea of all the integers as a totality. We appear to understand the idea of a totality of a dozen eggs, of a baker's dozen (thirteen) of rolls, 'and so on'. We then imagine that we understand the analogical idea of collecting *all* the integers into a set. Do we understand this clearly? Some thought not. Bernays gave the name platonism to the tendency to treat such collections as well understood, independently of how we come to understand them.

Bernays, like all his peers, grew up in a neo-Kantian milieu, and one where phenomenology was profoundly influential. The relation between the thinker and the thought was of cardinal importance, not as a matter of psychology but as fundamental to objective knowledge. Thus Bernays contrasts geometry as grounded by Euclid and as grounded by Hilbert, but not in the obvious superficial ways. 'Euclid speaks of figures to be *constructed*, whereas, for Hilbert, systems of points, straight lines, and planes exist from the outset' (1983: 259; these are the translator's italics, not those of Bernays).

> This example shows already that the tendency of which we are speaking consists in viewing the objects as cut off from all links with the reflecting subject.
>
> Since this tendency asserted itself especially in the philosophy of Plato, allow me to call it 'platonism'. (1983: 259)

To mimic my second, Hegelian, epigraph from Lakatos, Hilbert's points and lines have been 'alienated' from the reflecting subject who produces them. But it is not Hegelian alienation, not the creation of a new organism, the living, growing thing which is mathematics. It is cold, infinitely distant objects cut off from the reflecting subject.

In our days, when so many philosophers emphasize the embodied mind, one might protest that Euclid is constructing his diagrams not with the reflective mind but with ruler and compass, with hand and arm and pencil or stylus. I believe that is very important, yet at the same time there is much idealization. I don't mean the trite thought that the circle you inscribe is not a perfect geometric circle. Complex constructions are often too fraught with repeated error to work out at all. When I have tried actually to follow a construction of the regular 17-gon (§1.28), I never ended up with anything like an exact dissection of a circle into 17 equal segments. It was what Bernays called the reflective mind that did the construction, not the careless hand.

Bernays first explains platonism in terms of arithmetic. 'The weakest of the "platonistic" assumptions introduced by arithmetic is that of the totality of whole numbers' (1983: 259).

That is 'a *restricted platonism* which does not claim to be more than, so to speak, an ideal projection of a domain of thought' (1983: 261, italics added). But, he immediately continued,

> Several mathematicians and philosophers interpret the methods of platonism in the sense of conceptual realism, postulating the existence of a world of ideal objects containing all the objects and relations of pure mathematics. It is this absolute platonism which has been shown untenable by the antinomies.

Absolute platonism, asserting the existence of all definable mathematical objects and relations, is untenable. What remain are relative platonisms. The weakest interesting platonism described by Bernays asserts the existence of a totality of whole numbers.

After Bernays had introduced the word 'platonism' to the philosophy of mathematics, it should have been clear that one should not speak of platonism, but about platonism restricted to some domain of objects, such as the class of integers, or any Zermelo–Fraenkel set, or whatever. Perhaps even something like the class of all ZF sets, in a von Neumann–Gödel–Bernays set theory.

5   Other totalities

I have chosen a moment in history, the moment when 'platonism' became established in philosophical thinking about mathematics. The focus on totalities may itself be deemed virtually obsolete. It is true that most expositions of intuitionism or of constructive proofs do not emphasize totalities, but often that is what is lurking behind more recent discussion.

I shall take just one example. George Boolos (1940–96)* asked 'Must we believe in set theory?' He asked us to think about a large but not exorbitant cardinal κ. This cardinal 'is quite small, indeed *teensy*, by the standards of those who study large cardinals'. 'But it's a *pretty big* number, by the lights of those with no previous exposure to set theory, so big, it seems to me, that it calls into question the truth of any theory one of whose assertions is the claim that there are at least κ objects' (Boolos 1998: 120).

Boolos was wondering whether there are *that* many things, but I shall not pursue his elegant discussion. He also turned to an axiom familiar in the first week of an introductory course on set theory, the power set axiom: 'it does not seem to me unreasonable to think that perhaps it is not the case that for every set, there is a set of all its subsets' (Boolos 1998: 131). He ends by quoting a 1951 lecture of Gödel's (1995: 306), urging that our understanding of sets – the 'iterative' conception of Boolos (1971) – forces items such as the power set axiom on us. (The metaphor of forcing occurs also in Gödel (1983). Gödel was speaking of things like the power set axiom, not e.g. the axiom of choice.) Boolos leads us on to very much the kind of experience of the reflecting subject – he speaks often of reflection – that Bernays may have had in mind; but not in the framework of phenomenology.

In Bernays' vocabulary Boolos is a cautious platonist. He has no problem about the totality of whole numbers, but he has many qualms about sets whose existence is proven within Zermelo–Fraenkel set theory with the axiom of choice. But his qualms have nothing to do with contemporary nominalism/ platonism debates that were going on in his lifetime. These, as we shall see in Part B, typically involve the denial of 'abstract' objects. Boolos jeers at those who deny the existence of numbers. Of course we do not interact with numbers in the way we interact with stars, by some physical process. So what?

> We twentieth-century city dwellers deal with abstract objects *all the time*. We note with horror our *bank balances*. We listen to *radio programs*: *All Things Considered* is an abstract object. We read or write *reviews* of *books* and are depressed by *newspaper articles*. Some of us write *pieces of software*. Some of us compose *poems* or *palindromes*. We correct *mistakes*. We draw *triangles* in the sand or on the board . . .
>
> It is very much a philosopher's view that the only objects there are are physical or material objects, or regions of space-time, or whatever it is that philosophers tell us . . . To maintain that there aren't any numbers at all because numbers are abstract and not physical objects seems like a demented way to show respect for physics, which of course everyone admires. (Boolos 1998: 128f)

6   Arithmetical and geometrical totalities

I have attended only to that aspect of Bernays' paper of least philosophical interest. I have wanted only to recall the context in which the label 'platonism' was launched – and to insist that what he called platonism arose in a context rather different from that of contemporary platonism/ nominalism debates. It is often unclear what the point of platonism is. In Bernays the point of the various platonisms is absolutely clear. What totalities must be assumed in order to arrive at various chunks of interesting mathematics? For a fully classical mathematician, the sky is not the limit, but maybe heaven is. Bernays in effect draws out the main lines of modern proof theory, in which one examines what principles of argument, and what totalities, must be assumed in order to derive a given mathematical result.

Bernays' own mathematical philosophy, in 1934, is well indicated by this statement: 'the two tendencies, intuitionist and platonist, are both necessary; they complement each other, and it would be doing oneself violence to renounce one or the other' (1983: 269). Two *tendencies*. That's a bit like saying two *attitudes*, but, I think, stronger.

What suffices to show that the two tendencies are complementary? The intuitionist tendency befits arithmetic, and the platonist tendency befits geometry. And now things get complicated. 'The idea of the continuum is a geometrical idea which analysis expresses in terms of arithmetic' (1983: 268). We are returned to the astonishment of Chapter 1, (§§1.4ff) at how arithmetic and geometry seem to turn out, over and over again, to be 'the same stuff'.

Bernays remains loyal to Kant. He does *not* think they are the same stuff. He does speak of the 'duality of arithmetic and geometry' and says it is 'not unrelated to the opposition between intuitionism and platonism' (1983: 269). But these dualities and oppositions do *not* form 'a perfect symmetry'. For 'the concept of number appears in arithmetic. It is of intuitive origin, but then the idea of the totality of numbers is superimposed'. But on the other hand 'in geometry the platonistic idea of space is primordial' (*primordiale*).

The thought here plainly derives from Kant. We grasp the idea of number from repetition in time, which gives us only an ongoing series, never completed. We grasp the idea of space from lines and shapes and solids, completed. In his 1918 book on the continuum, Weyl had attempted an essentially intuitionist approach to the continuum which Bernays found unsatisfactory. Bernays was basically satisfied with the analysis of points on the continuum by the real numbers, as defined, ultimately arithmetically, by Dedekind cuts. But because of the way in which we apprehend time and space in fundamentally different ways, and because of the different ways in which the 'reflecting subject' comes to the idea of these as objective entities, they cannot be regarded as 'the same stuff'.

7   Then and now: different philosophical concerns

There is a ready illustration of the striking difference between Bernays' platonism/intuitionism and twenty-first-century platonism/nominalism. Bernays writes: 'A misunderstanding about Kronecker must first be dissipated, which could arise from his often-cited aphorism that the whole numbers were created by God, whereas everything else in mathematics is the work of man. If that were really Kronecker's opinion, he ought to admit the concept of the totality of whole numbers' (Bernays 1983: 262). But he rejected that concept. So he could not have uttered the aphorism.

Harold Edwards (b. 1936), a lifelong advocate of Kronecker and the constructivism that ensued, has written many helpful essays about him, including both constructivist completions of some of Kronecker's research programmes, and also a biographical appreciation (Edwards 1987). Like Bernays, he doubts that Kronecker could have meant what the aphorism seems to say; in fact for quite some time he thought that the aphorism is 'bogus' (emails of July 2012).

An old-fashioned nominalist would react very differently from Edwards and Bernays. If God did create the positive integers, then he created a sequence of individual numbers. But he did not create the set of all those numbers. After all, sets do not exist. Not even sets of individuals exist: only the individuals.

Such a thought never crossed Bernays' mind. I doubt that it crossed Kronecker's mind either.

Conversely, a twenty-first-century platonist will say: positive integers are abstract objects. Kronecker said that God created them. So he thought they exist. That's a species of platonism – about numbers. He was a very modest platonist, but a platonist all the same.

Such a thought seems never to have crossed Bernays' mind, for he thought of platonism in terms of totalities rather than 'abstract objects'. Bernays' platonism/intuitionism takes place in a discourse completely different from today's platonism/nominalism.

A brief discussion of the aphorism will lead us to yet another discourse. The opposed attitudes of Kronecker and Dedekind are very different from those professed by Connes and Gowers. I shall suggest what Kronecker *did* mean, not as an exercise in history or interpretation, but to bring out the force of these two earlier attitudes. I read Kronecker's aphorism not so much as a statement of belief as a sharp, if metaphorical, way of expressing a fundamental disagreement with Dedekind.

Where does the aphorism come from? Not from Kronecker's writings or *Nachlass*. It's from an obituary by a good friend and colleague, Heinrich Weber. He states that the saying will be well known to his readers. It comes, he tells us, from a talk Kronecker gave to the Berlin Naturforsher-Versammlung in 1886, in the form, *Der ganzen Zahlen hat der lieber Gott gemacht, alles anderes ist Menschenwerk* (Weber 1893: 15; for variants, see Ewald 1996: 942). The lecture is not preserved; it was apparently a talk about elliptic functions, a major concern of Kronecker's from 1883 on. Note that Weber was reporting five or six years after the event; oral history, as we know, is tricky.

The saying can be translated as, 'The good Lord made the whole numbers, but everything else is the work of man.'

8   Two more mathematicians, Kronecker and Dedekind

I shall return to what the aphorism means, but this is another opportunity to think about the very many possible philosophical *attitudes* to mathematical reality. Those of well over a century ago will seldom be identical to any attitudes expressed in the twenty-first century. The world has moved on. You *cannot* have the same attitude to the Folies Bergère (or to dancing girls) that contemporaries of Manet had when he painted the woman behind the bar in 1882; you cannot have the same attitudes to mathematics that philosophically minded German mathematicians had in the same decade.

Leopold Kronecker (1823–91) and Richard Dedekind (1831–1916) were near contemporaries. Both were heirs of Gauss, not only in that they grew up with the immense legacy of mathematics created by the master, but also as heirs of his philosophy, including that bit sketched in §5.9. Pure mathematics is arithmetic and, thereby, so is algebra; geometry is synthetic. But there is much more philosophy than that. In an essay on 'The concept of number' (1887) Kronecker explicitly attributes to Gauss the doctrine that 'Number, is *solely* the product of our mind' (quoted by Stein: 1988, 243; Kronecker there cites a letter by Gauss, 9 April 1830).

Thus the analyticity and aprioricity of arithmetic and algebra – independence from how the world is – are due to the fact that they are about objects of thought, 'creations' of the human mind. Not your mind or my mind, but 'the' mind. This idea is transmitted by Dirichlet, who builds in the notion, much exploited by Dedekind, that we *create* the numbers: see quotation (4) in the next section.

These mathematicians might be said to share a philosophical psychology, even a 'theory of mind', that is rather alien to the twenty-first century. They all take for granted what may be familiar to those who know Boole (1815–64), that Logic is the study of the Laws of Thought, which in turn are the laws of the human mind. For many years past, most analytic philosophers would regard it as polite to pass over such 'psychologism' – if not to denounce it. It is not nearly so alien to philosophers with a neo-Kantian and phenomenological background, such as Bernays.

Frege trashed Dedekind for psychologism, and Michael Dummett (1991: 49–52) heaped injury upon insult, calling Dedekind's work 'misbegotten' and Frege's refutation 'conclusive' (1991: 80). Tait (1996) defends Dedekind ably enough on this score. Kit Fine (1998) has a comparable defence of related matters in Cantor, whom Dedekind thanked enthusiastically for several expositions. That's enough to indicate matters are not simple, even within the self-confident world of analytic philosophy of mathematics.

9 Some things Dedekind said

Let us briefly explore Dedekind's attitude, remembering that nominalism and more recent platonism were simply not on his dance card. To repeat, the aim is only to point at various attitudes, so different from those of Gowers vs Connes, and so different from twenty-first-century agonizing about nominalism. I write in the spirit of someone touring a museum of antiquities, and hence will rely entirely on secondary material that quotes primary material. Because much of it sounds strange to a modern analytic ear, we know we are in the presence of a somewhat different discourse with different connotations and assumptions. Hence the words that I shall quote below have been interpreted in many ways. Mostly I prefer just to look at them: but I shall propose a self-conscious interpretation of Kronecker's aphorism in the light of them.

I shall number some of Dedekind's statements. José Ferreirós (1999, ch. VII) begins with two epigrams quoting Dedekind. One is from the famous *Was sind und was sollen die Zahlen* of 1888.

(1) Nothing capable of proof ought to be accepted in science without proof.

This is not as trite as it sounds. He is saying that the most elementary truths of arithmetic must be proved. That involves axioms. Many authors prefer to refer not to the Peano axioms but to the Dedekind–Peano axioms for arithmetic. But proofs are not just axioms; they are derivations which require Logic, the Laws of Thought.

Dedekind's own epigram for *Was sind* will be recalled from §5.11:

> (2) Man always arithmetizes.

Ferreirós follows (1) with a second epigram, which, whatever be your opinion of it, is certainly not trite. It is an extraordinary claim about Human Nature, perhaps an explication of (2):

> (3) Of all the aids which the human mind has for simplifying its life, i.e. the work in which thinking consists, none is so rich in consequences and so inseparably bound up with its innermost nature as the concept of number ... every thinking man [*jeder denkende Mensch*], even if he does not clearly realize, is a man of numbers, an arithmetician. (Ferreirós 1999: 215, quoting from an undated manuscript)

Finally, a word about Dedekind's invention of set theory. He used the word 'system' where we now use 'set'. Ferreirós (1999: 226) writes, with his italics:

> (4) He began by defining a '*thing*' (*Ding*) to be any object of our thought ... The notion of set of was explained as follows ... 'It very frequently happens that different things *a, b, c,* ... considered for any reason under a common point of view, are collected together in the mind, and one then says that they form a *System S*; one calls the things *a, b, c,* ... the *elements* of the system *S.*'

Like everyone else at the time, Dedekind took what we would now call a universal comprehension axiom for granted: any 'things' that can be considered, for any reason, under a common point of view, can be collected together into a totality, a system or set. Obviously that aspect of his approach needed revision.

The other scholar from whom I lift some quotations is Stein (1988). Dedekind was writing to Heinrich Weber (a good friend of both Dedekind and Kronecker). Weber had suggested a 'Russellian' definition of the numbers, as classes of classes similar to a given class. This takes for granted the concept of a 1:1 mapping, and as Stein and Ferreirós insist, Dedekind put the idea of mappings on a firm and fully general footing for the first time. Indeed, Dedekind believed that the ability to 'map' was the fundamental human capacity that makes mathematics possible.

Dedekind replied to Weber in friendly spirit, concluding with a remark about what we now call the Dedekind cut, the now standard tool for defining real numbers.

> (5) I should still advise that by number ... there be understood not the *class* (the system of all mutually similar finite systems), but rather something *new* (corresponding to this class) which the mind *creates*. We are of divine species and without doubt possess creative power not merely in material things

(railroads, telegraphs), but quite specially in intellectual things ... you say that the irrational number is nothing else than the cut itself, whereas I prefer to create something *new* (different from the cut), which corresponds to the cut. (Stein 1988: 248; cf. alternative translation in Ewald 1996: 835)

The opinion, that a real number is not *identical* to a certain sequence of rationals, sounds very much in the same spirit as Gowers, saying it would be foolish to say that complex numbers just *are* ordered pairs, or that a whole number *is* some set or other. What is interesting here is the notion of creation. *We are of divine species*: some hyperbole, that!

A couple of other aspects of Dedekind's philosophy should be mentioned. He thought of the ordinal numbers as primary, in contrast to those who began by defining the cardinal numbers. Second, he had a radical thought about the whole numbers. Most everyone thinks we start with a concept of whole number, and build up extensions of the number concept, perhaps in the way that Hamilton did (§4.25). Now Dedekind well knew and profited from Hamilton (1837), but he was deeply impressed by the characterization of simply infinite sets as those that can be mapped into some of their subsets, and thought the characterization of number might begin there.

10 What was Kronecker protesting?

Since profound admirers of Kronecker, such as Bernays and Edwards, have trouble with the aphorism about God creating the whole numbers, we might consider two options. First, he was misreported by his friend Weber, at a time when many of Weber's readers would actually have been present at Kronecker's 1886 talk five or six years earlier. More plausible is that Kronecker really did say something like the famous aphorism. So what can he have meant?

Perhaps this. He and Dedekind had the same philosophical and mathematical background, the same heroes, and the same projects. I suggest he pretty much agreed to each of (1)–(5) above, with one critical footnote. We do *not* create the whole numbers! They are given, if you like a gripping metaphor, by the good Lord. His attitude, at just this point, is that of Alain Connes and 'archaic reality'. But Connes has no problem with our creating 'projections' which we use to create organization within the blunt reality of the whole numbers, no problem with as much set theory as we care for. In that respect Connes resembles Dedekind.

Kronecker, in contrast, thought that the only way properly to comprehend mathematical reality was in terms of the numbers as we find them.

The free creations of the human mind are admirable. He is said to have congratulated Lindemann for proving that π is transcendental, i.e. not the root of any algebraic equation: great work but in the realm of human creation, not reality. (For more on this anecdote, however, see the final essay in Edwards (2005).)

The foundations of mathematics must, Kronecker thought, start with foundations, and cannot reside in our creations. Bernays (1983: 263 n. 2) observed how Kronecker himself developed the algebraic numbers entirely out of the integers. Edwards has continued the programme in a number of ways. We now see constructive mathematics as the continuation of a programme envisaged by Kronecker. But we may have forgotten that it comes out of a conception of mathematics and the mind shared with Dedekind, with the one critical difference.

To repeat: the foundations of mathematics cannot reside in our creations. That (I suggest) is why Kronecker said that the good Lord made the whole numbers, but the rest is the work of man. Kronecker thought there is one mathematical reality, that of the whole numbers, and all truth-seeking mathematics must be built up from that.

Dedekind thought that we, of divine species, create the mathematical reality; the whole numbers, the sets of whole numbers, and if need be the entire Cantorian superstructure, are all part of that reality which is our construction.

One can read Connes' Platonism as a curious blend of Dedekind and Kronecker. Something is not our creation, namely the series of natural numbers. 'Archaic' matches 'God-made'. After that, everything is our creation, and we project it on to the primitive reality. But that we create it does not mean it is mere fiction, no more than Dedekind's railroads are fictions. We are creators of a reality, which we use to investigate the archaic reality. There are no Kroneckerian inhibitions about what we can find out using the projections.

11   The structuralisms of mathematicians and philosophers distinguished

One of the most vigorous current philosophies of mathematics is named structuralism, but it has only a loose connection with the mathematician's structuralism of which we have spoken. Among its proponents one finds Resnik (1981, 1988), Hellman (1989, 2001), Parsons (1990, 2008), Shapiro (1997, 2000), and Chihara (2004, 2010). I introduce it here because Dedekind is often cited as its predecessor, with Benacerraf (1965) as its immediate instigator. Benacerraf (§6.12) observed that numbers cannot be identical to any one of their analyses in set theory, for those analyses are not identical. At best the integers can be the shared structure of all sound analyses. Stein (1988: 248) noticed that many of Dedekind's observations about identifying numbers with constructions look similar to those advanced by Benacerraf.

Dedekind did not use the label 'structuralist'. Bourbaki certainly did (§2.12). Many later French mathematicians such as André Lichnerowicz are content with it. To use a phrase I lifted from Erich Reck (2003: 371), their attitude is in 'a separate category, or of a different kind from that of the recent philosophers' who are concerned with 'semantic and metaphysical issues'. No labels fit well, so let us call structuralism of the Bourbaki type *mathematician's structuralism*, and that of recent analytic philosophy *philosopher's structuralism*.

To illustrate, a philosopher may say things such as, 'Structures exist whether they are exemplified in a nonstructural realm or not … In mathematics, anyway, the places of mathematical structures are as bona fide as any objects are' (Shapiro 1997: 89). A structuralist mathematician, in contrast, may describe himself as 'radically non-ontological' (Lichnerowicz in CL&S 2001: 25). Philosopher's structuralism is thoroughly ontological, even when it is nominalist, with a sparse cast of beings.

Chihara's structuralism is avowedly nominalist (2004, 2010). Hellman's title, *Mathematics without Numbers* (1989), makes plain its nominalist bent, though his version invokes a notion of logical necessity, which is hardly to the taste of Quine's classic nominalism. Hellman (2001) distinguishes types of structuralism. Reck (2003: 370–4) conveniently summarizes types of structuralism in three pages. He then presents a strand in each type of contemporary structuralism that can be seen as matching an idea of Dedekind's, but then points out other strands that are discordant with Dedekind. He proposes what he calls 'logical structuralism' to capture Dedekind's ideas in the discourse of present-day analytic philosophy.

Among philosophical structuralists, it appears that more of them avow nominalist leanings than favour platonistic tendencies. It is a standard objection, on which I shall not dwell, that many deny the existence of mathematical objects, but insist on the existence of mathematical structures – and those are the new objects of structuralist philosophy! *Plus ça change*, a platonism in new clothing. There is a companion doctrine in the philosophy of science, named structural realism (inaugurated by Worrall (1989); for a survey, see Ladyman (2009)). There, the 'realism' is avowed; one does not assert the existence of unobservable theoretical entities, but does assert the reality of structures. By analogy, 'structural realism' in the philosophy of mathematics would not assert the existence of numbers, but would assert the existence of numerical structures.

There is a sort of highest common factor of all these philosophical -isms. The following statement oversimplifies, but it may be a helpful template. *The structuralism of the philosophers always takes for granted standard denotational semantics (§6.22), while the structuralism of the mathematicians never does.* This is an instance of a more general phenomenon. Today's nominalism/platonism debates among philosophers are typically conducted within the framework of denotational semantics. So let us take a brief and superficial glance at those debates.

B   TODAY'S PLATONISM/NOMINALISM

12   Disclaimer

It could be argued that the central issue of analytic philosophy of mathematics these past thirty years involves versions of nominalism against versions of platonism, versions very different from the type of platonism that led Bernays to coin the latter word in the first place. For reasons elaborated in the next section, I do not believe they have much to do with mathematical activity, with proofs, or with the actual work of applied mathematicians, and hence do not fall under the topics that I announced at the beginning of this book (proof, application, and other mathematical activities). Hence I shall take no committed position on these issues. But (as mentioned once before) critics often complain that I never indicate where I locate myself on philosophical spectra of interest. Hence I shall rather briefly set out how these issues appear to me, and where I might locate myself – despite the fact that my personal location is of zero philosophical significance. Anyone who wants a serious examination of the contemporary platonist/nominalism debates in the philosophy of mathematics should, as suggested in §1 above, begin with the relevant articles in the *Stanford Encyclopedia of Philosophy*.

13   A brief history of nominalism now

John Burgess and Gideon Rosen begin their invaluable survey of *Strategies for a Nominalistic Interpretation of Mathematics* (1997) by describing a sequence of three periods of twentieth-century philosophical thinking about maths. First comes turmoil in foundations early in the century. They mean in part the 'crises', and not just the Russellian paradoxes, but more generally the questions about totalities to which Bernays directed us. 'Philosophy of mathematics then entered [they write] a period of comparative quiescence that lasted through the middle decades of the century.'

> Nominalism (as understood in contemporary philosophy of mathematics) arose toward the mid-century when philosophy of mathematics was otherwise rather quiet. It arose not in the mathematical community, as a response to developments within mathematics itself, but rather among philosophers, and to this day is motivated largely by the difficulty of fitting orthodox mathematics into a general philosophical account of the nature of knowledge. The difficulty largely arises from the fact that the special, 'abstract' objects apparently assumed to exist by orthodox mathematics – numbers, functions, sets, and their ilk – are so very different from ordinary, 'concrete' objects. Nominalism denies the existence of any such objects. (Burgess and Rosen 1997: vii)

They note that 'nominalism is distinctively a philosopher's concern', and report that 'the field of philosophy of mathematics today tends to be regarded among professional philosophers as a legitimate if not especially prestigious area of specialization, and among professional mathematicians as a dubious though probably harmless hobby' (1997: vii). That is my excuse for being perfunctory. Lots of mathematicians care about philosophy. Witness the translation and publication of *Triangle of Thoughts* (CL&S 2001) by the American Mathematical Society. Witness Connes and Gowers, not to mention Grothendieck and Voevodsky. But issues of nominalism, which at present preoccupy academic philosophers, do not much matter to mathematicians or to mathematics. They have very little to do with mathematical activity. That is why my glance at these topics will be fleeting.

In passing, let it be said that John Burgess is not a nominalist, and began to explain why in 1983, in one of a series of exceptionally clear papers reprinted in Burgess (2008).

For seven sections to follow I shall adopt the scare quotes of the above quotation, and speak of *'abstract' objects* rather than of *abstract objects*. When I return in §21 to contemporary platonism, I shall delete the scare quotes.

14   The nominalist programme

'Nominalism denies the existence of any such objects' as 'numbers, functions, sets, and their ilk'. I shall call that *nominalism in mathematics*, but in this chapter often, for brevity, refer to it simply as nominalism. It has connections with mediaeval nominalism, but is distinctively contemporary.

In what follows I shall seldom consider 'abstract' objects in general, but limit the discussion to the specific case of numbers. Thus the core of nominalism in mathematics will be understood in this way: *it denies that numbers exist.*

In contemporary debate, *platonism in mathematics* is opposed to nominalism in mathematics. It asserts that numbers (and many other 'abstract' objects) exist. Obviously this is very different from platonism as opposed to intuitionism. Thus we have two distinct types of counter-platonism. The anti-Platonism of the mathematicians of Chapter 6 is something else again, for it neither asserts nor denies that there are numbers; it makes no sense of such claims.

Goodman and Quine (1947), following Russell himself (§19 below), made a valiant attempt to produce a systematic and formal nominalist theory – their 'Steps toward a constructive nominalism'. But Quine's regimentation of scientific knowledge just could not get rid of classes, even if, piecemeal, this or that use of classes could be explicated and therefore eliminated ('explication is elimination', as in §6.12). Thus he became what I called a wan platonist. Note in passing that he was not much troubled by classes being 'abstract' – which has now become the bête noire of nominalism. 'Where classes offend is not just on the score, so doubtfully offensive, of their abstractness' (Quine 1960: 266). That's from the exhilarating §55, 'Whither classes?' in *Word and Object.*

And there matters pretty much stood until the first major project of contemporary nominalism in mathematics. That's Hartry Field's *Science without Numbers: A Defence of Nominalism* (1980). It is still the unrivalled benchmark of the programme. The title has been mimicked by others in various ways, always trying to do without numbers. A direct example comes from the structuralist Geoffrey Hellman, in *Mathematics without Numbers* (1989). I said in §12 above that nominalism/platonism did not address 'the actual work of applied mathematicians', but that does not mean Field and those who came after were not interested in application, for the indispensability argument, which much moved them, relies on the fact that mathematics is essential for the sciences. I meant only that the nominalists seldom attend to what applied mathematicians do. SIAM, we might say, has not crossed their path. And, unlike Quine, contemporary nominalists do find abstractness very 'offensive', for they cannot see how abstract objects can hook up with the material world.

An antidote to abstract-object loathing might be George Boolos (1998: 129), who teasingly asked, 'When may we expect *Computer Science Without Programs* to appear?' – programmes being just as much abstract objects as numbers. To recall Boolos from §5 above, he was also amusingly dismissive of worries about 'abstract' objects. He called the thesis that numbers do not exist 'demented'.

15 Why deny?

Why on earth would a scientist or mathematician want to do without numbers? The short answer is that none do. Many mathematicians are fascinated by the integers, while others devote their lives to the reals. The whole numbers are Connes' archaic reality; they are, in Kronecker's metaphor, God's gift to human beings. It would never occur to anti-Platonists such as Lichnerowicz or Gowers to try to engage in mathematics or science without numbers. It is true they don't *assert* that numbers exist, for that, in their view, makes no sense; likewise it makes no sense to deny that numbers exist.

I may mention a strikingly unusual but coherent opinion from the third figure on the *Triangle of Thoughts*, Paul Schützenberger. Rather than trying to do science without numbers, he holds that integers exist *only* in physics: 'in my opinion, outside of physics, there is no such thing as an integer; consequently, we are engaged only in empty mathematics' (CL&S 2001: 90). He calls this a Pythagorean thesis. What impresses him is that, for example, in crystallography there are certain intrinsic whole number relationships. That's where integers live. And if there is an integer that has no place in the world, then it does not exist. Hence it is very likely that many integers do not exist. In its context, an 'internal' context, this statement is on a par with the South African recorded voice telling me that the number I have dialled does not exist (§6.18).

So why do some philosophers want to do mathematics or science without numbers? First, a clarification. Field does not want to *do* science without numbers, only to show that we *could* dispense with numbers. But why would anyone want to? Let us examine the roots of nominalism in mathematics. They are, in my opinion, entwined with denotational semantics, as sketched in §6.22. But it is a little complicated. We might speak not of roots of nominalism, but of routes to nominalism, for paths diverge and join.

One route takes up the issues raised by Benacerraf (1973). A denotational semantics, for the most ordinary statements making use of numbers, implies that numbers exist. Analytic epistemologists today usually urge that you can know about an object only if you have some sort of causal or historical relationship to the object in question. It is obscure how we can have causal or historical relationships to 'abstract' objects such as numbers. That is why

Burgess and Rosen, as quoted above, spoke of 'the difficulty of fitting orthodox mathematics into a general philosophical account of the nature of knowledge'. Many readers now appear to assume that Benacerraf was offering an argument against platonism. To repeat what was said in §6.22, he presented a dilemma, with difficulties for each horn of the dilemma.

The other route follows Quine's indispensability argument, sketched in §19 below, and which involves our understanding of first-order quantifiers.

16   Russellian roots

Scholastic nominalists believed that individuals exist, but not much else. The universe does not have individuals *plus* collections of individuals, classes, universals, or other merely intellectual paraphernalia, creatures of the mind. I have said (§6.4) that Russell and Quine were born to be nominalists in something like that sense, even if the hard knocks of mathematical and philosophical experience shattered childhood complacency. (Russell actually began as an idealist in the Hegelian manner of late Victorian England, but that was his infancy, not his childhood.) Quine became an ontological relativist – how you talk indicates what you must believe to exist. And you should talk in the ways best suited to current scientific knowledge. That is the 'naturalized epistemology' which has become standard lore in present-day analytic philosophy.

Nominalist Russell, constitutionally inclined to say that classes do not exist, tried to develop what he called a no-class theory, closely connected with what he called 'the supreme maxim in scientific philosophizing: *Whenever possible, logical constructions are to be substituted for inferred entities*' (1917: 155, original italics; cf. Russell 1924: 386). After describing his logical construction of cardinal numbers, he asserted that 'a similar method, as I have shown elsewhere, can be applied to classes themselves, which need not be assumed to have metaphysical reality, but can be regarded as symbolically constructed fictions' (1917: 156).

The best short account of the complex moves involved, and their failings, is in the *Stanford Encyclopedia* article on logical construction by Bernard Linsky (2009). Scholars still debate whether Russell's attempts to avoid classes succeeded: Kremer (2008) vs Soames (2008). I personally do not think he did, but the argument continues. But it all began earlier, with Russell's 'On denoting' (1905).

He took a silly sentence, 'The present King of France is bald', whose subject expression is the definite description, 'The King of France'. The sentence is true if and only if there is a present K of F, there is at most one

K of F, and everything that is now a K of F is bald. Thus the definite description, which Russell called a denoting phrase, and which would now be called a singular term, could be eliminated. That is to say, statements about the present King of France could be restated as a conjunction of three statements none of which contains the denoting phrase, 'the present King of France'. There is only the supposedly unproblematic predicate, '*a* present king of France'. What makes the three-part conjunction false is the conjunct that says, 'There is a present K of F.'

No one seems to have noticed Russell's little joke, which must have been obvious in 1905. It goes all the way back to Richard Whately (1787–1863). The present King of France comes from the fourth, and standard, edition of *Elements of Logic* (Whately 1831). Now forgotten, this was the most widely used logic textbook in the English-speaking world, from its appearance in the *Encyclopaedia Metropolitana* in 1826, through most of the rest of the century. Max Fisch (1982: xv) reports that C. S. Peirce often told how he became obsessed with logic: when he was twelve (i.e. in 1871), he picked up Whately's textbook in his brother's rooms at Harvard, and never looked back. One finds the main parts of the book published (often together with examination questions) at least into the 1870s: for example, Whately (1874). Here is Whatley's present King of France: 'suppose we are speaking of the present King of France; he must actually *be* either at Paris or elsewhere; sitting, standing, or in some other posture; and in such and such a dress, *&c.*' (1831: 128, italics in original). Here is Russell's joke, now lost in the sands of time: even in 1831 there was no King of France! Louis XVIII reigned 1814–24, and Charles X 1824–30. The words may have been written in 1826, but remained in print for decades and decades during which there was no such king.

To return from schoolboy humour to logic, elimination is the name of the contemporary nominalist game. Eliminate, from scientific language, all singular terms that denote 'abstract' objects, and, in particular, numbers. Hence Hartry Field's *Science without Numbers: A Defence of Nominalism*.

Russell showed, as Quine succinctly put it, that 'whatever we say with the help of names can be said in a language which shuns names altogether' (Quine 1953: 13). The point is general. Quine wrote a fascinating introductory note to a selection from *Principia Mathematica* (Introduction, Chapter III, 'Incomplete symbols', §1 'Descriptions'). He faithfully summarized a method for using the theory of descriptions to eliminate names for classes and much else.

> More generally, Whitehead and Russell's way of eliminating descriptions enables us to make a clean sweep not only of class abstracts but of all other

sorts of singular terms as well, except variables. Instead of assuming a
constant singular term or proper name, say '*a*', we can always assume or
define a predicate '*A*' that is true only of the object *a*, and then take '*a*' itself as
short for the eliminable description '(*ιx)Ax*'. [The object *x* such that *Ax*.]
(Quine 1967: 217)

So in a regimented language, we eliminate all singular terms. There is no
problem about what they denote, because none remain.

17   Ontological commitment

But Quine had moved on from Russell. In that great paper of 1948, 'On
what there is' (Quine 1953), he coined the wonderful expression, 'ontological commitment', together with the aphorism, 'To be is to be the value of a
variable'. This is not to be used for adjudicating among rival ontologies, but
'in testing the conformity of a given remark or doctrine to a prior ontological standard. We look to bound variables in connection with ontology
not in order to know what there is, but in order to know what a given
remark or doctrine, ours or someone else's, *says* there is' (1953: 15).

These thoughts, together with what has come to be called his indispensability argument, led Quine to what I have somewhat impolitely called a
wan platonism (§6.4). Why wan? Because it does not commit one to much.
It does not feel like anything that would have moved Plato, or, for that
matter, Alain Connes.

'On what there is' did stamp a new direction on analytic philosophy. The
first four pages examine how one can coherently express disagreements
about what kinds of entities there are. On page 5 we encounter Russell's
'theory of so-called singular descriptions', which serves Quine's purposes for
a while, but then he makes the cardinal move of the paper. The '*only* way we
can involve ourselves in ontological commitments' is 'by our use of bound
variables' (1953: 12, original italics).

> The variables of quantification, 'something', 'nothing', 'everything', range
> over our whole ontology, whatever it may be; and we are convicted of a
> particular *ontological presupposition* if, and only if, the alleged presupposition
> has to be reckoned among the entities over which our variables range in order
> to render one of our affirmations true. (1953: 13, my italics)

Although almost all his readers speak of 'ontological commitment' – what a
gripping phrase! – I find the words 'ontological presupposition' more
satisfactory, even if it has less of Quine's panache with language. This is
not a matter of mathematics, but I shall briefly explain my preference.

18 Commitment

Commitment is in the province of moral philosophy. And also of theology, for faith speaks much of commitment, precisely because it is far stronger than mere belief. I commit myself to a cause. Marriage is a commitment, expressed, traditionally, in the words 'I do'.

If I make a statement to the immigration authorities, a declaration to the police, or other formal enunciation, especially if under oath, it is natural to say that I am committing myself. I put myself at risk in my declaration, for if what I say is not true, I have perjured myself, and might end up in jail.

If I am speaking as a professional, with the authority derived from my professional status – medicine and the law have long been the prototypical professions – my obligations, derived from my status, approach that of making an oath. Every type of authority exercised in context has, by its status, such obligations: scientist, teacher, dentist, or journalist. We recognize that diplomats and politicians are less committed to the truth of what they say than we would like them to be. We clearly hold experimental scientists communicating results to a high standard of commitment. If what they say they did, they did not do, or if what they say happened, did not happen, they are exposed, shamed, and even fired. More modestly, if I am telling you something, imparting information – the best route to the Forum, say – it is not unnatural to say that I am (or should be!) committed to the truth of what I say.

My own profession is teacher of philosophy, which is a high responsibility. Suppose I tell my class, by way of example, that the fifth perfect number is 3350336. I write it on the board. To what am I committed by that statement? Nothing much, I am afraid. A student raises her hand and says, 'There's something wrong, that looks too small'. I look and agree. I have left out a '5'. The fifth perfect number is 33550336. I apologize to the class. My commitment amounts to this – I should apologize if I got it wrong.

I have not even done the calculation to check that 33550336 is the sum of its proper divisors, I just looked it up online. But I can check the calculation with a hand calculator. 'Are you sure it is the *fifth* perfect number? Might you not have missed a smaller perfect number larger than 8128?' Not so easy! Even with these relatively small numbers, I am going to appeal to authority, probably online. That is how I back up my commitment.

'But at least you are committed to the existence of some perfect numbers?' Funny question, but yes, because 6 and 28 are obviously perfect, and it is easy enough to check 8128. 'So you are committed to the existence of numbers?' Now I have a problem. For what on earth could I do, to back up this commitment? What could I do to fail to honour it?

> A commitment that cannot be backed up, or not honoured, is no commitment at all.

That is why I believe it is better to speak of ontological presuppositions.

19   The indispensability argument

In Quine's opinion (as noted in §3.30), really 'pure' mathematics, with no known 'application' to the world about us, hardly deserves to be called true at all. Statements about perfect numbers look pretty 'pure' to me; I cannot think of an application, until I recall that some scholastics studied them in the hope of understanding perfection itself (§5.17, *App 6*). In Quine's view, generalizations about perfect numbers are not strictly true or false at all. They are merely provable, or not. So we need not worry what the statements are about, or whether names for perfect numbers denote anything at all.

But the natural sciences constantly invoke numbers, to the extent that William Thomson, Lord Kelvin (1824–1907), notoriously opined that if you can't measure something you know nothing about it, yet. More importantly all theoretical science involves numerical equations and more complicated functions. But that's not all: applied mathematics quantifies over numbers, functions, and their ilk. Hence, in Quine's ontology it presupposes their existence.

All this is not peculiar to the sciences, although physics is the favourite haunt of analytic philosophers discussing questions of formal ontology. At the time of the 2012 Olympics, millions of peoples were talking about Usain Bolt, 'the fastest man ever'. Try saying he is the fastest man ever without quantifying over numbers. Try saying without quantifying over numbers that the world record for the 100 metres is 9.63 seconds, namely, that *that* number of hundredths of a second is less than any number of seconds in which any person other than Bolt has run 100 metres.

Does a conversation about Usain Bolt really presuppose that numbers exist? Get real. His record does presuppose an enormous amount of technology, time-pieces, cameras, and the whole paraphernalia of modern racing, including the laying out of the track to exactly 100 metres, with all lanes of a uniform distance. Without that, there would be no 9.63 seconds on the record books. I am a little inclined to pursue the previous reflections on 'commitment' with 'presupposition'. Time-pieces and the apparatus of measuring are presupposed in a very different way from that in which numbers are said to be presupposed.

Be that as it may, here is the classic brief statement of the indispensability argument:

quantification over mathematical entities is indispensable for science, both formal and physical; therefore we should accept such quantification; but this commits us to the existence of the mathematical entities in question. This type of argument stems, of course, from Quine, who has for years stressed both the indispensability of quantification over mathematical entities and the intellectual dishonesty of denying the existence of what one daily presupposes. (Putnam 1979: 347; cf. 1971)

The argument is often called the Quine–Putnam argument, but unlike Quine, Putnam did not repeat it much. He began a talk at a 1990 conference honouring Quine by listing some of Quine's views, including the indispensability argument, and by saying that he had shared them ever since he 'was a student at Harvard in 1948–9' [the year of 'On what there is']. 'But, I must confess they are views that I now want to criticize' (Putnam 1994: 246). He went on to say that his views in 1990 had 'features in common with the philosophy of the later Wittgenstein rather than with that of Carnap'.

Putnam's rethinking about Wittgenstein has progressed over a long period, is very important to philosophy, and has to be read in its own right. But we might put in a word for Carnap here: his distinction between internal and external assertions of existence. It needs more careful work, but the Carnapian response to indispensability arguments would be that the only existence statements presupposed by quantifying over numbers are purely formal and external. They have no 'ontological' import.

Liggins (2008) argues on textual grounds that neither Putnam nor Quine *ever* advanced the indispensability argument as now understood. That is a historical, biographical, and interpretive question that I shall not address, except to recall the splendid aphorism already cited in §6.12, 'explication is elimination' (Quine 1960: 263). You can explicate numbers, Quine urged, in many a way, and thereby eliminate them. But classes are recalcitrant. His 'holism' has been so emphasized by his readers that few notice how he so often prefers to proceed piecemeal. 'Scope still remains for nominalism, however, and for various intermediate grades of abnegation of abstract objects, when with Conant we think of science not as one evolving world view but as a multiplicity of working theories' (Quine 1960: 270; the references is to Conant 1952: 98ff).

> The nominalist can realize his predilection in special branches, and point with pride to a theoretical improvement of those branches. In the same spirit even the mathematician, realist *ex officio*, is always glad to find that some particular mathematical results that had been thought to depend on functions or classes of numbers, for instance, can be proved anew without appealing to objects other than numbers. (Quine 1960: 270)

The mathematician is *ex officio* a platonist!

To return to Putnam's statement quoted above, whatever he intended (which I leave to Liggins and others), he did display antagonism to '*denying* the existence of what one daily *presupposes*'. There are two verbs here, and hence two questions. (1) In what sense does one presuppose the existence of numbers when one is engaged in the formal or physical sciences? (2) If one declines to deny, does one thereby have to assert?

Evidently the anti-Platonists of Chapter 6 intend neither to deny the existence of numbers nor to assert it. Their answer, and I believe the correct answer, to question (2) is a firm 'No!' From that point of view, the indispensability argument counts *against* nominalism in mathematics. But it does not thereby, in itself, *favour* any species of platonism or Platonism. I cannot imagine mathematicians of Platonist inclinations using the indispensability argument to back up their convictions. It is irrelevant to mathematics, except in the sense that a mathematician could (conceivably) embark on Field's programme of showing how to eliminate quantification over mathematical entities in scientific practice. Not, I suggest, a promising career path for a young person seeking tenure in a department of mathematics.

So we turn to question (1), about presupposition.

20 Presupposition

For a good four decades (bringing us up to 1988) Quine's indispensability argument appears to have received no concerted challenge. Slowly, grumblings of a very high calibre began to appear (e.g. Sober 1993; Melia 1995, 2000; Maddy 1997, 2007; Yablo 1998; Azzouni 2004). They take different tacks. Penelope Maddy, strongly professing a naturalistic philosophy, calls into question Quine's celebrated holism. A physical science demands mathematics, but we do not have to treat every statement in physics + mathematics as on a par with every other one. The language of science is more structured than that, and admits of parts with different types of ontologies. The existential claims and presuppositions in one type of work may be very different from others. Since I myself have argued for the disunities of the sciences (both in the plural) I inevitably sympathize, though on broader grounds, perhaps, than Maddy's (Hacking 1996).

Jody Azzouni, Joseph Melia, and Steve Yablo question, in various ways, simplistic understandings of existence. From a distance, they connect with Carnap's instinct for distinguishing internal and external statements, but their analyses are far more refined, as befits the passage of time.

Against this work, Mark Colyvan, author of the informative article on indispensability in the *Stanford Encyclopedia of Philosophy* (2011), argues (2010) that 'There is no easy road to nominalism'. I believe he means in part that subtle work on quantification does not get over the basic problem of the first route to nominalism, namely denotational semantics, deriving from Benacerraf (1973).

I think that he is right. You need a more radical approach, and of course mine would start from a version of Wittgenstein's maxim (§6.23), don't ask for the denotation (*Bedeutung*), ask for the use. Among all critics of the indispensability argument known to me, the closest to that maxim is Thomas Hofweber (2006, 2007). The title of the 2007 paper (whose publication was delayed; it may have been written first) is 'Innocent statements and their metaphysically loaded counterparts'. 'Jupiter has 67 confirmed moons' is innocent, while '67 is the number of confirmed moons of Jupiter' is metaphysically loaded – at least in denotational semantics. Hofweber wants to unload the metaphysics from the latter statement.

Hofweber makes no use of Wittgenstein's maxim, so I asked him to what extent he agreed with it, and also with the anti-semantic opinion of Chomsky, cited in §6.24. I quote with permission from an email of 29 August 2012.

> I think I don't fully agree with either Chomsky or Wittgenstein in the quotes you give, but I think I am congenial to them in spirit. I do believe that there is a useful project of a semantics for a natural language, but such a project can't be separated from, and is in a sense secondary to, how expressions are used by the speakers of the language. For a number of the cases that I have looked at, number words, for example, the important question is what speakers are doing when they use number words in their speech, and here a crucial distinction is whether they use them with the intention to refer to objects, or otherwise. The semantics of the language speakers use is then derivative on that, and so in a sense use is before *Bedeutung*.

I take no well-informed stance on semantics, but my ill-informed attitude welcomes this programme of research. Hofweber also wrote in the same mail that one goal is

> to find out what we are doing when we talk about numbers. Here there is a puzzling feature that such talk has, and we need to understand why talk about numbers exhibits this puzzling behavior, in particular why number words appear as both determiners and as singular terms, and why they appear to be used as names for objects in some uses, but not in others. To understand our puzzling talk about numbers is the goal, and of course to see what conclusions we can draw for the philosophy of arithmetic from all this.

It is an obvious complaint about the Wittgensteinian maxim (ask for the use), that it invites waffle. In Hofweber's work we finally find someone getting down to the nitty-gritty of 'use', and being willing to produce generalizations, perhaps even encoded in a semantic theory, as opposed to a string of examples.

I do not wish to imply that Hofweber aligns himself with anything Wittgenstein said, over and above his stated 'congeniality in spirit' to one famous sentence. But I believe that anyone who does feel aligned – as I do – with many aspects of Wittgenstein's thinking, will find Hofweber's research far more congenial than most recent writing about the indispensability argument and the uses of number words.

21   Contemporary platonism in mathematics

There are numerous explicit and detailed avowals of nominalism, only a few of which have been cited above. But perhaps because it is a default position, the same cannot be said for platonism. As regards mathematics, the number of platonists many times exceeds the number of nominalists, but philosophical expositions of nominalism far exceed philosophical expositions of platonism. Hence it is good to have James Robert Brown's *Philosophy of Mathematics* (2008).* It is exemplary of a Victorian genre regrettably rare today: both a textbook for novices and an impassioned argument for philosophical ideas. It poses the question, 'What is Platonism?' on page 12, and answers by stating seven platonist tenets in the course of three pages. I strongly commend these for clarity and brevity. There is no shilly-shallying. Here is the first proposition, with original italics:

> *Mathematical objects are perfectly real and exist independently of us.*

Note the hyper-adjective: *perfectly* real. Are some objects imperfectly real? Hyper-adjectives, such as we first noticed in Kant (§3.19) are gesticulations, expressing the feeling that the plain statement does not say it all, and yet there is no more that can be said. Beware. Brown goes on to say that we discover these objects. 'Any well-formed sentence of mathematics is true or false, and what makes it so are the objects to which it refers.'

Brown immediately makes plain the connection with denotational semantics. 'Platonism and standard semantics (as it is often called) go hand in hand.' Brown's example: suppose that Mary loves ice cream. Brown uses semantic ascent, inviting us to suppose that

> the sentence 'Mary loves ice cream' is true. What makes it so? In answering such a question, we'd say 'Mary' refers to the person Mary, 'ice cream' to the substance, and 'loves' refers to a particular relation which holds between

Mary and ice cream. It follows rather trivially from this that Mary exists. (2008: 13)

I find this excellent because it is so up-front. It allows me once again to practise pedantry. Notice how far 'refers' has moved from ordinary usage. *We'd say.* We = those who practise 'standard' referential (denotational) semantics. My instinct is to recall Strawson 'On referring' (§5.24). You won't find me saying that 'loves' refers to a relation. If I said something about Mary, and was asked, 'Which Mary?', I might reply: 'I was referring to Mary Poppins, not Mary Magdalene'. (And it does not follow at all that Mary Poppins exists!) I can't think of an occasion when I would say that I used the verb 'to love' to refer to anything. As Strawson insisted, expressions do not refer, but we use expressions to refer – in my case, to Mary Poppins.

Brown moves on to the sentence '$7 + 5 = 12$'. With the same chain of thoughts as in the passage just quoted, it is said to follow rather trivially, that '=' refers to a relation that holds between the sum $7 + 5$ and the number 12. Hence the number 12 exists.

It is the great merit of Brown's exposition that the centrality of the 'standard' semantical account is put up front, in lay terms, for all to see. In a footnote he refers to the full argumentation in Bob Hale's book *Abstract Objects* (1987).

Bernays rightly urged that any platonism must be relative. You can no longer say that mathematical objects are perfectly real; on pain of contradiction you have to clarify the objects you have in mind. In conversation Brown agreed that he ought to have taken advice from Bernays. He said that when he wrote of mathematical objects, he intended to include Zermelo-Fraenkel sets. In an email of 14 June 2012, quoted with permission, he wrote, however:

> I am no longer the fan of set theory that I was. Platonists who come out of set theory are influenced by arguments about truth, reference and the like. These are okay, but the kind of thing that impresses me most are intuitions. This is the thing that puts most people off platonism; they find such epistemology to be absurd. But the fact that so many mathematicians can have intuitions that are so solid and remarkable, suggests that there is something like the perception of an independent reality going on.

Indeed his fourth tenet, in its original italics, is: '*We can intuit mathematical objects and grasp mathematical truths.*' 'Mathematical entities can be "seen" or "grasped" with "the mind's eye".' Here Brown might avail himself of the very careful examination of intuition by Charles Parsons (2008, ch. 5).

22   Intuition

Right from the start, §1.30, I have stayed away from 'intuition', for reasons partially explained there. In at least Western philosophy, sight has always been the prime metaphor for knowledge, and there are good reasons for its being so (Ayers 1991). And there is certainly insight. What about the sort of mathematical entities Brown is discussing? One of them is the·number 27. I can handle 27 pretty well, I can see at a glance that it is the cube of three and can check that in a moment of mental arithmetic. But I would not describe myself as seeing or grasping the number 27 with my mind's eye. Would you? There are people who see numbers in colours; that is a variety of what is called synaesthesia, but it is held to be an unusual neurological condition. Really good number theorists can usefully be described as having a personal acquaintance with many interesting numbers, and some will support this with visual metaphors. Well and good. But many do not describe themselves that way at all; it does not seem essential to being good at number theory.

I have stayed away from intuition as a term of art, but *of course* we have intuitions. And *of course* mathematicians have intuitions about, for example, numbers or knots or non-commutative algebra. We have the experience of suspecting something without being able to state a good reason for it. We also know that some people's hunches are as a rule better than others. So, lacking other knowledge, we may trust them. For example, when it comes to judgements of character, Judith's intuitions are far better than mine. She can tell a newly met phoney in five seconds, but I can't. In mathematics, such feelings play an important role in exploring a topic. They suggest conjectures, and they spur research programmes. When we have to decide quickly about a course of action, a hunch may be all we have to go on. But intuitions, in *this* sense, are never good grounds for firm beliefs, unless inductively: she's usually right about people; Euler was usually right about numbers.

On the other hand, recall Descartes' phrase from §1.30: 'an indubitable conception formed by an unclouded and attentive mind; one that originates solely from the light of reason'. That's a pretty good description of what some mathematicians strive for. And they may well talk about it in terms of intuitions. Well and good, but as I have said, I decline to follow suit.

Brown is right: in recent times many mathematicians speak about intuitions. It would be an interesting question for a historical sociologist of mathematics to address: when did talk of intuition become commonplace in mathematics? For one instance of the use of the word, see the final quotation in this chapter, which also serves as the last word in this book.

Intuitions in mathematics are not simply 'had'; they are cultivated. They are taught in the course of mathematical explanation or teaching. A nice example is furnished by the current *abc* conjecture. It jumped into prominence in September 2012 with a proof put online by Shinichi Mochizuki in late August 2012 (Ball 2012). The conjecture emerged only in the 1980s, but turns out to have many powerful consequences for number theory. It is in the same bag of facts and conjectures about primes as Fermat's last theorem: it states that there are only finitely many triples of positive integers that satisfy a certain condition involving prime numbers. Barry Mazur (n.d.) has a beautiful exposition of what it is all about. After setting out some basics, he writes, 'But perhaps it is time to backtrack, to develop a bit of intuition, which might allow us to hazard guesses on which equations . . . could be expected to have few solutions, and which could be expected to have many' (n.d.: 8). In fact, on skimming the various blogs about Mochizuki's work, it is remarkable how many speak of developing intuitions in this way.

At the time of writing, most people say that Mochizuki's proof is extraordinarily difficult, very long – 500 pages – and introduces new objects that are not familiar to the mathematical community. It is thought that if it works it will be revolutionary. I expect that it will produce a host of interesting philosophical reflections. Some have deep intuitions about it. They know how to think about it. But that is a far cry from Brown's assertion that mathematicians intuit mathematical objects. Beware of slippage between the same words in different uses!

23 What's the point of platonism?

I find Brown's change of gear, away from denotations, and towards intuition, very instructive. What really moves Brown (2008) is his fascination with pictorial proofs. He is a champion of pictorial argument, and his subtitles speak of *The World of Proofs and Pictures*. Quite possibly platonism provides the best story about how pictorial arguments can be so persuasive. The metaphors of sight are perhaps inevitable.

What I personally like about Brown's platonism is that here the idea does some work. To my mind, much contemporary platonism in mathematics fails to explain anything about mathematical practice. Brown offers a platonistic explanation of picture proofs. Perhaps it even offers an account of cartesian proofs (§1.19), the proofs one can get in one's mind as a whole. But it does not seem to me to explain anything about the typical proof one may explore in a mathematics seminar this afternoon, let alone in the published article one may read this evening. But that is a personal reaction: others will find platonism of this sort enormously helpful.

At any rate one may contrast Brown's platonism with another variety, due to Balaguer (1998). He surveys a great many arguments for and against platonism in mathematics, and finds most of them inconclusive. He himself favours what he calls 'full-blooded' platonism, by which he means that all possible abstract objects exist.

There are many possible quibbles, starting with the question of whether the idea is coherent. The quibbles can degenerate into silliness. Recall Boolos in §5 above, suggesting that *radio programme* is an abstract object, and also that a particular long-running programme on US National Public Radio, *All Things Considered*, is an abstract object. Now start to contemplate all possible radio programmes – do all these possible objects exist?

Leaving aside such quibbles, I have no sense at all of what it would be like for full-blooded platonism to be true or false. What would it be like for only some consistent abstract objects to exist, or none of them? I cannot see what difference Balaguer's full-blooded platonism makes. It seems to me to be bloodless.

24   Peirce: The only kind of thinking that has ever advanced human culture

Perhaps we are not seeing the wood for the trees, not seeing the big picture because we focus on so many small ones. I conclude these two chapters with a big picture taken from Charles Sanders Peirce. Unfortunately he used a big and uncomfortable word to express it, which he once spelt 'hypostatization' and once 'hypostatisation'. Present-day dictionaries are uncomfortable with it; the *Shorter Oxford* omits it. *Baldwin's Dictionary of Philosophy and Psychology* (1902) is contemporary with Peirce; in fact he contributed many of the entries, but not this one, due to J. M. Baldwin himself: 'The verb to hypostasize (sometimes written hypostasise) is used of the making actual or counting real abstract conceptions.' I shall use the 'z' throughout, even when quoting a passage that uses the 's'.

This is followed by an unsigned account of the origins of the word and its cognates in third- and fourth-century Christian theology (to which one could add Plotinus). It ends: 'The history of the matter is still far from being clearly understood.' At least some historians of theology would repeat that sentence today. Whatever word we use, and whatever its history, that *making abstract conceptions actual, or counting them as real,* may capture what is going on in the past two chapters.

Numbers are abstract conceptions; Platonism makes them actual or counts them as real. Nominalism, both scholastic and contemporary, rejects that gambit. Perhaps the idea of hypostatization offers a new way to think. The platonist says numbers *are* real. The hypostasizer (to coin a new and ugly word) counts them as real, or makes them actual. To hypostatize is to do something: it is thought in action. Peirce thought our ability to do so is one of our greatest assets. To judge by context, this 'intuition' is probably intended to recall Kant:

> Intuition is the regarding of the abstract in a concrete form, by the realistic hypostatization of relations; that is the one sole method of valuable thought. Very shallow is the prevalent notion that this is something to be avoided. You might as well say at once that reasoning is to be avoided because it has led to so much error; quite in the same philistine line of thought would that be; and so well in accord with the spirit of nominalism that I wonder some one [*sic*] does not put it forward. The true precept is not to abstain from hypostatization, but to do it intelligently. (Peirce *CP* 1: 383 (1890))

The manuscript breaks off at this point. It is part of a draft MS of a book to be titled *A Guess at the Riddle*. No hint as to how we should hypostatize intelligently survives. A few years later we find:

> It may be said that mathematical reasoning (which is the only deductive reasoning, if not absolutely, at least eminently) almost entirely turns on the consideration of abstractions as if they were objects. The protest of nominalism against such hypostatization . . . as it was and is formulated, is simply a protest against the only kind of thinking that has ever advanced human culture. (*CP* 3: 509 (1897))

This is from a long paper, 'The logic of relatives' (relations) published in *The Monist*. For the importance of this and similar pieces, see Hilary Putnam on Peirce the logician (1982). The dots of omission stand for rude remarks about 'the empty disputation of medieval Dunces'. I shall not try, here, to elucidate Peirce's thought about hypostasis. A further grasp of his idea can be had from his numerous remarks about abstraction. I wish only to register it as a way of looking at the forest rather than the trees. What a brash statement – 'the only kind of thinking that has advanced human culture'.

It should be said that our family of languages – traditionally called Indo-European – lends hypostasis more than a helping hand. We use sentences in which names for any old object, 'abstract' or 'concrete', serve as grammatical subjects. This leads on to a point emphasized by Nietzsche long ago: European languages demand an existential presupposition for terms in the subject position. European grammars generate hypostasis. Here is another philosophical polarity too seldom noticed. Nietzsche thought this created endless bad philosophy. Peirce had to agree in part ('medieval Dunces'), but maintained that it was precisely what had also advanced human culture to such an extent.

Peirce treats hypostatization as a matter of abstraction. I regard it as name magic, Wittgenstein's alchemy – recall from §6.22: 'Is it already mathematical alchemy, that mathematical propositions are regarded as statements about mathematical objects, – and mathematics as the exploration of these objects?' Maybe we do need alchemy, name magic, whatever derogatory epithet you choose, in order to advance human culture.

25   Where do I stand on today's platonism/nominalism?

Obviously, I do not think it is important. I would happily be diagnosed as an Ockhamite. To repeat from §6.18, Lichnerowicz's Ockham thought that 'the realists and nominalists are fighting over nothing' (CL&S 2001: 37). His Ockham sounds a bit like my resurrected and reconstructed scholastic nominalist (end of §6.7). We can tax wagons; we can talk about taxing wagons; we do not need wagonhood to do so. We use numbers all the time. More rarefied souls are fascinated by them, study their properties. But we do not need abstract objects. Stop asking for the denotation, ask for the use.

To lack interest in a contemporary philosophical problematic is not to dismiss the philosophy of mathematics. I remain stuck on the very reasons that there is philosophy of mathematics at all. Thousands of authors have expressed the dilemma, but I shall simply requote Langlands; a more complete text is in §2.1: 'mathematics, and not only its basic concepts, exists independently of us. This is a notion that is hard to credit, but hard for a professional mathematician to do without.'

That is one of the reasons that the philosophy of mathematics is perennial.

I don't think we understand the enigma either: to requote from §1.2, 'the enigmatic matching of nature with mathematics and of mathematics by nature'. How come Robert Langlands' wholly 'pure' mathematics, in its geometrical version, could be picked up by Edward Witten, master of string theory, as an explication of the duality of electricity and magnetism in *this* universe? (§1.15) (Or is it any universe?)

26   The last word

I shall give the last word in these two long chapters in Plato's name to André Lichnerowicz, the formalist anti-Platonist of Chapter 6. He is telling us that when he is creative he is working with mathematical 'beings' – his word. He experiences them; he becomes familiar with them. They are not the 'abstract' objects of the philosophers, but vital realities. Yet at the end of the process, Lichnerowicz sees himself doing something different, namely producing a rigorous, formal proof. While he is at work, exploring, with the hope of creating new maths, he *experiences* himself as if he were a Platonist. But when he writes up the proof, all that is put behind him, in a formalism that at the same time has to be compelling.

This is no barren formalism. On the contrary, it is at the core of the applicability of mathematics. 'We have gained the ability to construct models, the ability to say that while engaged in hydraulics we can say exactly the same things we do in electrostatics, because we encounter the same equations in both fields. We have learned that when we do mathematics, Being with a capital "B" is disregarded ... This abstraction, a radical abstraction, is what gives mathematics both its theoretical power and its rich connections to reality.'

> Yet, it is not for these reasons, namely to produce rigorous and compelling proofs, that one becomes a mathematician. Every mathematician, within himself and even with a few close colleagues, conducts a discourse that has nothing to do with that. We must draw a clear distinction between the discourse of universal communication and the discourse of creation in mathematics. According to general consensus, when a mathematician works, he is in fact reflecting upon a certain field in which he encounters mathematical beings, and ends up playing with them, until they become familiar to him. He can then work while taking a walk or while conversing with a boring interlocutor. This is a somewhat draining activity which we have all experienced. After a time, be it a month or a year, either we get nowhere, finding nothing, or else we manage to get hold of some result. Then begins the onerous task – the obligation to write up for the benefit of the mathematical community a polished article that is as compelling as possible.
>
> As a result, there are two types of activity. If one becomes a mathematician it is for the creative activity, the game of intuition, and not at all for the burden of publishing. Too often the two are confused. (CL&S 2001: 24f)

Disclosures

These are brief notes on passages marked by a * in the text.

My choice of individuals, whose work is used as an example, is often influenced by some personal acquaintance or professional connection. I hope that does not matter to the argument, but here are some purely personal comments on the connections.

§1.15 *Robert Langlands* graduated from the University of British Columbia in 1957; I graduated in 1956. We were enrolled in a very small undergraduate programme in mathematics and physics, and so took a number of classes together. He was soon to become a permanent member of the Princeton Institute for Advanced Study. It is my own fault that I have not taken more advantage of this initial acquaintance with a major figure in twentieth-century mathematics.

§1.22 *Michael Harris*, of the Mathematical Institute at Jussieu (Paris VII), came to see me a number of times in Paris, and we had several conversations at the Fields symposium mentioned in §1.13. I owe to Harris both the mention of Vladimir Voevodsky (§1.22), and the snippets from Alexander Grothendieck (§1.26).

§1.27 *Imre Lakatos* and I arrived in Cambridge in the Michaelmas term 1956, he a seasoned revolutionary and I a gauche colonial. I came to know him fairly well in about 1958, and read much of his doctoral thesis (to which I devoted a whole chapter of Hacking 1962). I read the entire text of Lakatos (1963–4) in draft. (The version of 1976 includes more material which I did not see, including my third epigraph.) Sometimes I would venture to correct his English and he would quote Shakespeare back at me. The preface to the original published version acknowledges a number of people, including one Dr I. Hacking, who had barely received his doctorate. There is much more of Imre's influence in the present book than meets the eye.

§2.3   For Persian, thanks to Kaave Lajevardi. For Russian, thanks to Jane
        Frances Hacking. For Japanese, thanks to Yasuhiro Okazawa. For
        Chinese, thanks to Christian Wenzel of Taiwan National
        University.

§2.7   *J. L. Austin*. In 1957 (or thereabouts) Austin gave a talk at the
        Cambridge Moral Sciences Club, after having tea with a few
        undergraduates at Trinity College. At tea he dryly pontificated:
        'Moore and Frege were the only two careful philosophers.'

§2.12   This use of Lyotard's terminology is not to be understood as
        endorsing his views in any other way. Sometime after the
        publication of *La condition postmoderne: rapport sur le savoir* (Paris:
        Minuit, 1979) I was asked by Lindsay Waters, then at the University
        of Minneapolis Press, for my opinion on commissioning a
        translation. My report was utterly hostile. I was wrong: of course the
        book needed to be published in English.

§2.14   G. E. M. Anscombe even loaned me a couple of carbon copies of
        Wittgenstein's typescripts (the original carbons prepared by his
        typist). I can see why some of his stuff may have gone missing!

§2.18   The topologist was *Larry Siebenmann* of Orsay (Paris XI). He
        brought me Changeux and Connes (1989) as a Christmas present
        the year it was published, which is when I first thought about the
        passages quoted in §§3.7–8 and §3.15 – over a decade before I
        became acquainted with the authors.

§2.18   I have known *David Epstein* since 1956; he was the mathematician I
        knew best when I was an undergraduate at Cambridge. Hence the
        reference in his letter to 'when we were students'.

§2.21   *Stephen Cook* is a colleague at the University of Toronto (in the
        departments of mathematics and of computer science).

§3.17   I was recently surprised to see that the title of my review of Kitcher
        (Hacking 1984) was 'Where does math come from?' That could
        have been my title for the third Descartes Lecture. But it was the
        editor, Robert Silvers, who chose the title, not I. I recall (but it was a
        long time ago, and recollections are not reliable) Kitcher writing to
        me at the time saying the comparison with Mill was apt.

§3.20   *Casimir Lewy* is my own role model of a 'careful analytic
        philosopher'. For an obituary giving some flavour of his work and
        character, see Hacking (2006). *Wikipedia* describes my teacher as
        'Casimir Lewy, a former student of Wittgenstein's'. That is totally
        misleading. Yes, Lewy went to many of Wittgenstein's classes, and

Wittgenstein speaks of him in recorded notes, but Lewy's man was G. E. Moore, of whom he became the literary executor, and from whom he learned to be a 'careful analytic philosopher'.

§3.22 *Indefinable*?? In Russell's early philosophical understanding, one could not take the simple route of W. E. Johnston (1858–1931), who wrote that 'On my interpretation . . . the hypothetical "If *a*, then *b*" means the conjunction of *a* with *not-b* is false' (Johnson 1892: 18). In the first edition of *PM* in 1910, the difficulty has been resolved. 'The idea of implication, in the form in which we require it, can be defined' (*PM* 2nd edn: 44). Then follows the definition: [*1.01. *p* ⊃ q .=. ~pvq Df]. Negation and disjunction were primitives, replaced by Sheffer's stroke in the second edition.

§3.22 *'Necessity'* (Moore 1900). It comes as some surprise that in 1900 Moore was very dubious about necessity, and thought there were only degrees of material implication. Thomas Baldwin (email 10 September 2010) thinks he had changed his mind 'by 1903 since he gives fundamental truths concerning intrinsic value a special, absolute, status'. He regarded them as 'universal', applying not only to actual but also to possible beings. Moreover, 'in his unpublished review of Russell's *Principles* (written *c*. 1905) Moore argued that Russell should have distinguished entailment from implication'.

§3.32 Some oral history: Kripke (1965) presented his semantics for intuitionistic logic at a conference held in Oxford in 1963. The first intervention was by the influential youngish logician, Richard Montague (1930–71). 'This is an historic occasion. For the first time, intuitionistic logic has been made intelligible.' This is a profoundly *logicist* reaction to the fundamental contribution made by Kripke to intuitionistic logic. (It is also the only occasion when I have been present at what someone then rightly called an historic occasion!)

§4.17 *Reviel Netz* has been my invaluable guide through Greek proof; I do not pay nearly enough attention to his criticisms. On an unexpected topic, his *Barbed Wire: An Ecology of Modernity* (Wesleyan University Press, 2004) is a gripping book.

§5.10 I owe several of the more unusual references to Gergonne to Vincent Guillin (who first helped me locate the end of Albert Dadas in my book *Mad Travelers*).

§5.13 Hardy's *Pure Mathematics* – I've had my copy of the tenth edition (1952) since Christmas 1954.

§5.15   I doubtless attend more to Smith's Prizes than most. In 1960
        I shared the prize with Keith Moffatt. My essay was on modal logic,
        and his on magneto-hydrodynamic turbulence. He held a Chair in
        Mathematical Physics at Cambridge University, 1980–2002.

§5.16   I owe the information about von Mises to Friedrich Stadler,
        founder and director of the Institut Wienerkreis in Vienna.

§5.17   (*App 6*) *Chandler Davis* (b. 1926) is a close neighbour in Toronto.
        Among many other things he was a long-time editor of *The
        Mathematical Intelligencer*.

§5.20   *Weinberg*: I owe the reference to Aaron Sidney Wright, who is
        completing a doctorate on the vacuum at the University of Toronto.

§5.25   *Francis Everitt*, director of Gravity Probe B at Stanford, was a major
        informant for parts of my *Representing and Intervening* (1983), and
        my advisor on all matters Maxwellian.

§6.2   *Bill Tait* was a colleague when I briefly taught at the University of
        Illinois at Chicago Circle (as it was then called) under the
        chairmanship of Ruth Marcus.

§6.7   *Connes* and *Gowers* make a very good pair of mathematicians who
        disagree fundamentally about the nature of their subject, and so
        serve well the purposes of this chapter. I should disclose, however,
        that Connes is a colleague at the Collège de France, and Gowers is a
        Fellow of Trinity College, Cambridge, where I studied when young
        and of which I am an honorary Fellow. I enjoy the fact that the
        founders of the two institutions, François I (Collège de France,
        1530) and Henry VIII (Trinity, 1546), respectively aged twenty-six
        and twenty-nine, met at the Field of the Cloth of Gold (1520) to
        agree on a non-aggression treaty. (The two young men wrestled:
        François trounced Henry.)

§6.12   *Paul Benacerraf*. In 1960–61 I was employed as a 'preceptor' at
        Princeton where Paul was a young assistant professor. We talked a
        great deal about mathematics.

§7.5   *George Boolos*. In my year at Princeton, Benacerraf and I examined
        Boolos' senior thesis. It was clear then that he would go on to make
        major contributions to logic and the philosophy of mathematics.

§7.21   *James Robert Brown* is a colleague in the University of Toronto
        Department of Philosophy.

References

Almeida, Mauro W. and Barbosa, D. 1990. Symmetry and entropy: mathematical metaphors in the work of Lévi-Strauss. *Current Anthropology* 31: 367–85.

Ambrose, Alice. 1959. Proof and the theorem proved. *Mind* 68: 435–45.

Anderson, Philip. 1972. More is different: broken symmetry and the nature of the hierarchical structure of science. *Science* 177: 393–6.

Appel, K. and Haken, W. 1977. Every planar map is four colourable. Part I: Discharging. *Illinois Journal of Mathematics* 21: 429–90.

Appel, K., Haken, W., and Koch, J. 1977. Every planar map is four colourable. Part II: Reducibility. *Illinois Journal of Mathematics* 21: 491–567.

Aristotle. 1984. *The Complete Works of Aristotle*, revised Oxford translation, ed. Jonathan Barnes, 2 vols. Princeton University Press.

Aschbacher, Michael and Smith, Stephen D. 2004. *The Classification of Quasithin Groups*. Mathematical Surveys and Monographs vols. III, II2. Providence, RI: American Mathematical Society.

Asper, Markus. 2009. The two cultures of mathematics in ancient Greece. In E. Robson and J. Stedall (eds.) *Oxford Handbook of the History of Mathematics*. Oxford University Press, 107–32

Aspray, William and Kitcher, Philip. 1988. *History and Philosophy of Modern Mathematics*. Minneapolis: University of Minnesota Press.

Atiyah, Michael. 1984. An interview with Michael Atiyah. *The Mathematical Intelligencer* 6: 9–19.

  n.d. Sir Michael Atiyah on math, physics and fun. Available at: www.super stringtheory.com/people/atiyah.html.

Atiyah, Michael and Sutcliffe, Paul. 2003. Polyhedra in physics, chemistry and geometry. *Milan Journal of Mathematics* 71: 33–58.

Atiyah, Michael, Dijkgraaf, Robbert, and Hitchin, Nigel. 2010. Geometry and physics. *Philosophical Transactions of the Royal Society* 368: 913–26.

Austin, John Langshaw. 1961. The meaning of a word. In *Philosophical Papers*. Oxford: Clarendon Press, 23–43.

Ayer, A. J. 1946. *Language, Truth and Logic*, second edn, revised. London: Victor Gollancz; first edn 1936.

Ayers, Michael. 1991. *Locke*. London: Routledge.

Azzouni, Jody. 2004. *Deflating Existential Consequence: A Case for Nominalism*. Oxford University Press.

Bacon, Francis. 1857–74. *The Works of Francis Bacon*, ed. J. Spedding, R. Ellis, and D. Heath, 3 vols. London: Longman.

Baker, Alan. 2008. Experimental mathematics. *Erkenntnis* 68: 331–44.

Baker, G. P. and Hacker, P. M. S. 2005. *Wittgenstein. Understanding and Meaning: An Analytical Commentary on the Philosophical Investigations*, second extensively revised edn by Peter Hacker. Oxford: Blackwell; first edn 1980.

Balaguer, Mark. 1998. *Platonism and Anti-Platonism in Mathematics*. Oxford University Press.

Ball, Philip. 2012. Proof claimed for deep connection between primes. *Nature* 12 September. Available at: www.nature.com/news/proof-claimed-for-deep-connection-between-primes-1.11378.

Baron-Cohen, Simon. 2002. The extreme male brain theory of autism. *Trends in Cognitive Sciences* 6: 248–54.

2003. *The Essential Difference: Men, Women and the Extreme Male Brain*. London: Allen Lane.

Baron-Cohen, Simon, Wheelwright, S., Scott, C., Bolton, P., and Goodyear, I. 1997. Is there a link between engineering and autism? *Autism*, 1, 101–9.

Barzin, Marcel. 1935. Sur la crise contemporaine des mathématiques. *L'Enseignement mathématique* 34: 5–11.

Benacerraf, Paul. 1965. What numbers could not be. *Philosophical Review* 74: 47–73.

1967. God, the devil, and Gödel. *The Monist* 51: 9–32.

1973. Mathematical truth. *Journal of Philosophy* 70: 661–79.

Benacerraf, Paul and Putnam, Hilary. 1964. *Philosophy of Mathematics: Selected Readings*. Englewood Cliffs, NJ: Prentice-Hall.

1983. *Philosophy of Mathematics: Selected Readings*, second and much modified edn. Cambridge University Press.

Bernays, Paul. 1935a. Sur le platonisme mathématique. *L'Enseignement mathématique* 34: 52–69.

1935b. Quelques points essentiels de la métamathématique. *L'Enseignement mathématique* 34: 70–95.

1964. Comments on Ludwig Wittgenstein's *Remarks on the Foundations of Mathematics*, trans. from the German original. *Ratio* 2 (1959): 1–22, in Benacerraf and Putnam (1964: 510–28). See also the online Bernays Project, available at: www.phil.cmu.edu/projects/bernays/, text 23.

1983. On platonism in mathematics, trans. Charles Parsons of Bernays (1935a). In Benacerraf and Putnam (1983: 258–71). For a slightly revised translation, see the online Bernays Project, available at: www.phil.cmu.edu/projects/bernays/, text 1.

Blake, William. 1957. *The Complete Writings of William Blake*, ed. Geoffrey Keynes. London: Nonesuch Press.

Błaszczyk, Piotr, Katz, Mikhail, and Sherry, David. 2013. Ten misconceptions from the history of analysis and their debunking. *Foundations of Science* 18: 43–74.

Bloor, David. 2011. *The Enigma of the Aerofoil: Rival Theories in Aerodynamics 1909–1930*. Princeton University Press.

Boolos, George. 1971. The iterative conception of set. *Journal of Philosophy* 68: 215–31. Reprinted in Benacerraf and Putnam (1983: 486–502).

   1987. The consistency of Frege's *Foundations of Arithmetic*. In Richard Jeffrey (ed.) *Logic, Logic and Logic*. Cambridge, MA: Harvard University Press, 183–201.

   1998. Must we believe in set theory? In Richard Jeffrey (ed.) *Logic, Logic, and Logic*. Cambridge, MA: Harvard University Press, 120–31.

Borcherds, Richard. 2011. Renormalization and quantum field theory. *Algebra and Number Theory* 5: 627–58.

Bourbaki, N. 1939. *Éléments de mathématique*. Paris: Hermann. There are nine postwar volumes of the project; a final fascicule, on commutative algebra, appeared in 1998.

Boyer, Carl B. 1991. *A History of Mathematics*, second edn, revised by Uta C. Merzbach. New York: Wiley.

Brooks, R. L., Smith, C. A. B., Stone, A. H., and Tutte, W. T. 1940. The dissection of rectangles into squares. *Duke Mathematical Journal* 7: 312–40.

Brown, Gary I. 1991. The evolution of the term 'Mixed Mathematics'. *Journal of the History of Ideas* 52: 81–102.

Brown, James Robert. 2008. *Philosophy of Mathematics. A Contemporary Introduction to the World of Proofs and Pictures*, second edn. London: Routledge.

   2012. *Platonism, Naturalism, and Mathematical Knowledge*. London: Routledge.

Burgess, John P. 2008. *Mathematics, Models, and Modality: Selected Philosophical Essays*. Cambridge University Press.

Burgess, John P. and Rosen, Gideon. 1997. *A Subject with no Object: Strategies for a Nominalistic Interpretation of Mathematics*. Oxford: Clarendon Press.

Burnyeat, Miles. 2000. Plato on why mathematics is good for the soul. In T. J. Smiley (ed.) *Mathematics and Necessity: Essays in the History of Philosophy*. Oxford University Press for the British Academy, 1–82.

Butler, Joseph. 1827. *Fifteen Sermons Preached at the Rolls Chapel*. Cambridge: Hilliard and Brown; Boston: Hilliard, Gray, Little, and Wilkins.

Buzaglo, Meir. 2002. *The Logic of Concept Expansion*. Cambridge University Press.

Carey, Susan. 2009. *The Origin of Concepts*. Oxford University Press.

Carnap, Rudolf. 1950. Empiricism, semantics, and ontology. *Revue Internationale de Philosophie* 4: 20–40. Reprinted in Benacerraf and Putnam (1983: 241–57).

Cartwright, Nancy. 1983. *How the Laws of Physics Lie*. Oxford University Press.

Cauchy, Augustin-Louis. 1821. *Cours d'analyse de l'École Polytechnique*. Paris: Imprimerie Royale.

Chemla, Karine and Guo Shuchun. 2004. *Les neuf chapitres: le classique mathématique de la Chine ancienne et ses commentaires*. Paris: Dunod.

Changeux, Jean-Pierre and Connes, Alain. 1995. *Conversations on Mind, Matter, and Mathematics*, Princeton University Press. French original, *Matière à penser*. Paris: Odile Jacob, 1989.

Chevreul, Eugène. 1816. Recherches chimiques sur les corps gras, et particulièrement sur leurs combinaisons avec les alcalis. Sixième mémoire – examen des graisses d'homme, de mouton, de bœuf, de jaguar et d'oie. *Annales de Chimie et de Physique* 2: 339–72.

Chihara, Charles. 2004. *A Structural Account of Mathematics*. Oxford University Press.

2010. New directions for nominalist philosophers of mathematics. *Synthese* 176: 153–75.

Chomsky, Noam. 1966. *Cartesian Linguistics: A Chapter in the History of Rationalist Thought*. New York: Harper and Row.

2000. *New Horizons in the Study of Language and Mind*. Cambridge University Press.

Cipra, Barry. 2002. Think and grow rich. In *What's Happening in the Mathematical Sciences*, vol. v. Washington: American Mathematical Society, 77–87.

Cohen, I. Bernard. 1985. *Revolution in Science*. Cambridge, MA: Harvard University Press.

Cohen, Paul J. 1966. *Set Theory and the Continuum Hypothesis*. New York: Benjamin.

Coleridge, Samuel Taylor. 1835. *Specimens of the Table Talk of Samuel Taylor Coleridge*, ed. Henry Nelson Coleridge, 2 vols. London: John Murray.

Colyvan, Mark. 2010. There is no easy road to nominalism. *Mind* 119: 285–306.

2011. Indispensability arguments in the philosophy of mathematics. In *Stanford Encyclopedia of Philosophy*, ed. Edward N. Zalta (spring 2011 edn). Available at: http://plato.stanford.edu/archives/spr2011/entries/mathphil-indis/.

Conant, J. B. 1952. *Modern Science and Modern Man*. New York: Columbia University Press.

Connelly, Robert. 1999. Tensegrity structures: why are they stable? In M. F. Thorpe and P. M. Duxbury (eds.) *Rigidity Theory and Applications*. New York: Kluwer Academic/Plenum, 47–54.

Connes, Alain. 2000. La réalité mathématique archaïque. *La Recherche* 332: 109.

2004. Repenser l'espace et la symétrie. *La Recherche* 381: 27.

Connes, Alain and Kreimer, Dirk. 2000. Renormalization in quantum field theory and the Riemann–Hilbert problem. 1. The Hopf algebra structure of graphs and the main theorem. *Communications in Mathematical Physics* 210: 249–73.

Connes, Alain, Lichnerowicz, André, and Schützenberger, Marcel Paul. 2001. *Triangle of Thoughts*. Providence, RI: American Mathematical Society. French original: *Triangle des pensées*. Paris: Odile Jacob, 2000.

Conway, John. 2001. *On Numbers and Games*, second edn. Natick, MA: A. K. Peters.

Conway, John, Burgiel, Heidi, and Goodman-Strauss, Chaim. 2008. *The Symmetries of Things*. Wellesley, MA: A. K. Peters.

Corfield, David. 2003. *Towards a Philosophy of Real Mathematics*. Cambridge University Press.

Corry, Leo. 2009. Writing the ultimate mathematical textbook: Nicolas Bourbaki's *Éléments de mathématique*. In E. Robson and J. Stedall (eds.) *Oxford Handbook of the History of Mathematics*. Oxford University Press, 565–87.

Courant, Richard and Robbins, Herbert. 1996. *What is Mathematics? An Elementary Approach to Ideas and Methods*, second edn, revised by Ian Stewart. Oxford University Press.

Crombie, A. C. 1994. *Styles of Scientific Thinking in the European Tradition: The History of Argument and Explanation Especially in the Mathematical and Biomedical Sciences and Arts*, 3 vols. London: Duckworth.

Dahan-Dalmedico, A. 1986. Un texte de philosophie mathématique de Gergonne (mémoire inédit déposé à l'Académie de Bordeaux). *Revue de l' Histoire des Sciences* 39: 97–126.

Daston, Lorraine. 1988. Fitting numbers to the world: the case of probability theory. In Aspray and Kitcher (1988: 221–37).

Davis, Chandler. 1994. Where did twentieth-century mathematics go wrong? In Sasaki Chikara, Sugiura Mitsuo, and Joseph W. Dauben (eds.) *The Intersection of History and Mathematics*. Boston: Birkhäuser, 120–42.

Davis, Martin. 2000. *The Universal Computer: The Road from Leibniz to Turing.* New York: Norton.

Dawson, John. 2008. The reception of Gödel's incompleteness theorem. In Thomas Drucker (ed.) *Perspectives on the History of Mathematical Logic.* Boston: Birkhäuser, 84–100.

Dedekind, Richard. 1996. What are numbers and what should they be? In Ewald (1996: 787–832). (Translation of *Was sind und was sollen die Zahlen?* Braunschweig: Vieweg, 1888.)

Dehaene, Stanislas. 1997. *The Number Sense: How the Mind Creates Mathematics.* London: Allen Lane/Penguin.

Dennett, Daniel. 2003. *Freedom Evolves.* New York: Penguin.

Derrida, Jacques. 1994. *Specters of Marx: The State of the Debt, the Work of Mourning, and the New International.* New York: Routledge. Original: *Spectres de Marx.* Paris: Galilée, 1993.

Descartes, René. 1954. *The Geometry of René Descartes*, trans. from the French of 1637 and Latin of 1649 by D. E. Smith and M. L. Latham. New York: Dover.
     1971. *Philosophical Writings*, ed. and trans. E. Anscombe and P. T. Geach. London: Nelson.

Detlefsen, Michael. 1998. Mathematics, foundations of. In E. Craig (ed.) *Routledge Encyclopedia of Philosophy*, 10 vols. London and New York: Routledge, vol. VI, 181–92.

Dirac, P. A. M. 1939. The relation between mathematics and physics. *Proceedings of the Royal Society (Edinburgh)* 59 (II): 122–29. Reprinted in *The Collected Works of P. A. M. Dirac, 1924–1948*, ed. R. H. Dalitz. Cambridge University Press, 1995, 907–14.

Douglas, Mary and Wildavsky, Aaron. 1982. *Risk and Culture: An Essay on the Selection of Technological and Environmental Dangers.* Berkeley: University of California Press.

Doxiadis, Apostolos and Mazur, Barry (eds.). 2012. *Circles Disturbed: The Interplay of Mathematics and Narrative.* Princeton University Press.

Dudeny, Henry Ernest. 1907. *The Canterbury Puzzles, and Other Curious Problems.* London: Nelson. Available online from Project Gutenberg.
     1917. *Amusements in Mathematics.* London: Nelson. Available online from Project Gutenberg.

Dummett, Michael. 1973. *Frege: Philosophy of Language*. London: Duckworth.

1978. *Truth and Other Enigmas*. London: Duckworth.

1980. Frege and analytical philosophy. *London Review of Books*, 18 September.

1991. *Frege: Philosophy of Mathematics*. London: Duckworth.

Edwards, Harold M. 1987. An appreciation of Kronecker. *Mathematical Intelligencer* 9: 28–35.

2005. *Essays in Constructive Mathematics*. New York: Springer.

Eckert, Michael. 2006. *The Dawn of Fluid Dynamics: A Discipline Between Science and Technology*. New York: Wiley.

Everitt, C. W. F. 1975. *James Clerk Maxwell: Physicist and Natural Philosopher*. New York: Scribner.

Ewald, William. 1996. *From Kant to Hilbert: Readings in the Foundations of Mathematics*, 2 vols. Oxford University Press.

Ferreirós, José. 1999. *Labyrinth of Thought: A History of Set Theory and its Role in Modern Mathematics*. Basel: Birkhäuser; revised edn with similar pagination, 2007.

2007. Ὁ Θεός Ἀριθμητίζει: The rise of pure mathematics as arithmetic with Gauss. In Goldstein, Schappacher, and Schwermer (2007: 235–68).

Field, Hartry. 1980. *Science without Numbers: A Defence of Nominalism*. Oxford: Blackwell.

Fine, Kit. 1998. Cantorian abstraction: A reconstruction and a defense. *Journal of Philosophy* 95: 599–634.

Fisch, Max. 1982. Introduction to Volume 1. In *Writings of Charles S. Peirce: A Chronological Edition*. Bloomington: Indiana University Press, xiv–xxv.

Fitzgerald, Michael. 2004. *Autism and Creativity: Is there a Link between Autism in Men and Exceptional Ability?* Hove, East Sussex: Brunner-Routledge.

2005. *The Genesis of Artistic Creativity: Asperger's Syndrome and the Arts*. London: Jessica Kingsley.

2009. *Attention Deficit Hyperactivity Disorder: Creativity, Novelty Seeking, and Risk*. New York: Nova.

Fitzgerald, Michael and James, Ioan. 2007. *The Mind of the Mathematician*. Baltimore, MD: Johns Hopkins University Press.

Fitzgerald, Michael and O'Brien, Brendan. 2007. *Genius Genes: How Asperger Talents Changed the World*. Shawnee Mission, KS: Autism Asperger Publishing Company.

Frege, Gottlob. 1884. *Die Grunlagen der Arithmetic. Eine logisch-mathematische Untersuchung über den Begriff der Zahl*. Breslau. *The Foundations of Arithmetic: A Logico-Mathematical Enquiry into the Concept of Number*, trans. J. L. Austin, second revised edn. Oxford: Blackwell, 1953.

1892. Über Sinn und Bedeutung. *Zeitschrift für Philosophie und philosophische Kritik* 100: 25–50.

1949. On sense and nominatum. Herbert Feigl's translation of Frege 1892. In H. Feigl and W. Sellars, *Readings in Philosophical Analysis*. New York: Appleton-Century-Croft, 85–102.

1952. *Translations from the Philosophical Writings of Gottlob Frege*, trans. Peter Geach and Max Black. Oxford: Blackwell; third revised edn. 1980.

1967. *The Basic Laws of Arithmetic: Exposition of the System*. Montgomery Furth's translation of selections from *Grundgesetze der Arithmetik*. Berkeley: University of California Press.

1997. *The Frege Reader*, ed. Michael Beaney. Oxford: Blackwell.

Friedman, Michael. 1992. *Kant and the Exact Sciences*. Cambridge, MA: Harvard University Press.

Gabbay, Dov (ed.). 1994. *What is a Logical System?* Oxford: Clarendon Press.

Galison, Peter. 1979. Minkowski's space-time: from visual thinking to the absolute world. *Historical Studies in the Physical Sciences* 10: 85–121.

Garber, Daniel. 1992. *Descartes' Metaphysical Physics*. University of Chicago Press.

Gardner, Martin. 1978. *Aha! Aha! Insight*. New York: Scientific American.

Gauss, Karl Friedrich. 1900. *Werke,* vol. VIII: *Arithmetik und Algebra*. Leibniz: Teubner; reproduced Hildesheim: Olms, 1981.

Gaukroger, Stephen. 1989. *Cartesian Logic: An Essay on Descartes' Conception of Inference*. Oxford University Press.

1995. *Descartes: An Intellectual Biography*. Oxford: Clarendon Press.

Gelbart, Stephen. 1984. An elementary introduction to the Langlands program. *Bulletin of the American Mathematical Society* 10: 177–219.

Gergonne, Joseph Diaz. 1810. Prospectus. *Annales de mathématiques pures et appliquées* 1: i–iv.

1818. Essai sur la théorie des définitions. *Annales de mathématiques pures et appliquées* 9: 1–35.

Gingras, Yves. 2001. What did mathematics do to physics? *History of Science* 39: 383–416.

Gödel, Kurt. 1983. What is Cantor's continuum problem? A revised version of the paper in *American Mathematical Monthly* 54 (1947): 515–25, prepared for Benacerraf and Putnam (1964) and reprinted in Benacerraf and Putnam (1983: 470–85).

1995. Some basic theorems on the foundations of mathematics and their implications. The Gibbs Lecture to the American Mathematical Society, 1951. In Solomon Feferman *et al.* (eds.) *Kurt Gödel: Collected Works*, 5 vols. Oxford: Clarendon Press, vol. III, 304–23.

Goldstein, Catherine, Schappacher, Norbert, and Schwermer, Joachim (eds.). 2007. *The Shaping of Arithmetic: After C. F. Gauss's Disquisitiones Arithmeticae*. Berlin: Springer.

Goodman, Nelson, and Quine, W. V. 1947. Steps toward a constructive nominalism. *Journal of Symbolic Logic* 12: 105–22.

Gowers, Timothy. n.d. The two cultures of mathematics. Available at: www.dpmms.cam.ac.uk/~wtg10/2cultures.pdf.

2002. *Mathematics: A Very Short Introduction*. Oxford University Press.

2006. Does mathematics need a philosophy? In Reuben Hersh (ed.) *18 Unconventional Essays on the Nature of Mathematics*. New York: Springer Science+Business Media, 182–201.

(ed.). 2008. *The Princeton Companion to Mathematics*. Princeton University Press.

2012. Vividness in mathematics and narrative. In Doxiadis and Mazur (2012: 211–31).

Grabiner, Judith. 1981. *The Origins of Cauchy's Rigorous Calculus*. Cambridge, MA: MIT Press.

Grattan-Guinness, Ivor. 2000. A sideways look at Hilbert's twenty-three problems of 1900. *Notices of the American Mathematical Society* 47: 752–7.

Griess, Robert. 1982. The friendly giant. *Inventiones Mathematicae* 69: 1–102.

Guéroult, Martial. 1953. *Descartes selon l'ordre des raisons*, 2 vols. Paris: Aubier.

Haas, Karlheinz. 2008. Carl Friedrich Hindenberg. In *Complete Dictionary of Scientific Biography*. Available at: www.encyclopedia.com/doc/1G2-2830902007.html.

Hacker, Andrew. 2012. Is algebra necessary? *New York Times Sunday Review*, 29 July.

Hacker, P. M. S. 1996. *Wittgenstein's Place in Twentieth-Century Analytic Philosophy*. Oxford: Blackwell.

Hacking, Ian MacDougall. 1962. Part I: Proof; Part II: Strict implication and natural deduction. Cambridge University PhD, 4160.

1973. Leibniz and Descartes: proof and eternal truth. *Proceedings of the British Academy* 59: 175–188. Reprinted in Hacking (2002: 200–13).

1974. Infinite analysis. *Studia Leibniz* 6: 126–30.

1979. What is logic? *Journal of Philosophy* 86: 285–319.

1982. Wittgenstein as philosophical psychologist. *New York Review of Books*, 1 April. Reprinted in Hacking (2002: 214–26).

1984. Where does math come from? *New York Review of Books*, 16 February.

1990. *The Taming of Chance*. Cambridge University Press.

1996. The disunities of the sciences. In P. Galison and D. J. Stump (eds.) *The Disunity of Science: Boundaries, Contexts and Power*. Stanford University Press, 37–74.

1999. *The Social Construction of What?* Cambridge, MA: Harvard University Press.

2000a. How inevitable are the results of successful science? *Philosophy of Science* 67: S58–S71.

2000b. What mathematics has done to some and only some philosophers. In T. J. Smiley (ed.) *Mathematics and Necessity: Essays in the History of Philosophy*. Oxford University Press for the British Academy, 83–138.

2002. *Historical Ontology*. Cambridge, MA: Harvard University Press.

2006. Casimir Lewy 1919–1991. *Proceedings of the British Academy* 138: 170–7.

2007. Trees of logic, trees of porphyry. In J. Heilbron (ed.) *Advancements of Learning: Essays in Honour of Paolo Rossi*. Florence: Olshki, 146–97.

2011a. Why is there philosophy of mathematics AT ALL? *South African Journal of Philosophy* 30: 1–15. Reprinted in Mircea Pitti (ed.) *Best Writing in Mathematics 2011*, Princeton University Press, 2012.

2011b. Wittgenstein, necessity and the application of mathematics. *South African Journal of Philosophy* 30: 80–92.

2012a. 'Language, truth and reason' thirty years later. *Studies in History and Philosophy of Science* 43: 599–609.

2012b. The lure of Pythagoras. *Iyyun: The Jerusalem Philosophical Quarterly* 61: 1–26.

2012c. Introductory essay. In T. S. Kuhn, *The Structure of Scientific Revolutions*, 50th anniversary edition. University of Chicago Press, vii–xxviii.

Hadot, Pierre. 2006. *The Veil of Isis: An Essay on the History of the Idea of Nature*. Cambridge, MA: Harvard University Press. Original: Paris: Gallimard, 2004.

Hale, Bob. 1987. *Abstract Objects*. Oxford: Blackwell.

Hamilton, William Rowan. 1833. Introductory lecture on astronomy. *Dublin University Review and Quarterly Magazine* 1: 72–85.

1837. Theory of conjugate functions, or algebraic concepts, with a preliminary and elementary essay on algebra as the science of pure time. *Transactions of the Royal Irish Academy* 17: 293–422.

1853. *Lectures on quaternions; containing a systematic statement of a new mathematical method; of which the principles were communicated in 1843 to the Royal Irish academy; &c*. Dublin: Hodges and Smith.

Harris, Michael. Forthcoming. *Not Merely Good, True and Beautiful*. Princeton University Press.

Hardy, G. H. 1908. *A Course of Pure Mathematics*. Cambridge University Press.

1929. Mathematical proof. *Mind* 38: 1–25.

1940. *A Mathematician's Apology*. Cambridge University Press.

Hardy, G. H. and Wright, E. M. 1938. *An Introduction to the Theory of Numbers*. Oxford: Clarendon Press.

Heilbron, John. 2010. *Galileo*. Oxford University Press.

Hellman, Geoffrey. 1989. *Mathematics without Numbers*. Oxford University Press.

2001. Three varieties of mathematical structuralism. *Philosophia Mathematica* 9: 184–211.

Hempel, C. G. 1945. On the nature of mathematical truth. *American Mathematical Monthly* 52: 543–56.

Herbrand, Jacques. 1967a. Investigations in proof theory: The properties of true propositions. In Van Heijenoort (1967: 525–81). French original 1930.

1967b. On the consistency of arithmetic. In Van Heijenoort (1967: 618–28). French original 1931.

Herodotus. 1972. *Histories*. London: Penguin Classics.

Hersh, Reuben. 1997. *What is Mathematics, Really?* Oxford University Press.

(ed.). 2006. *18 Unconventional Essays on the Nature of Mathematics*. New York: Springer Science+Business Media.

Hesse, Mary. 1966. *Models and Analogies in Science*. University of Notre Dame Press.

Hobbs, Arthur M. and Oxley, James G. 2004. William T. Tutte (1917–2002). *Notices of the American Mathematical Society* 51: 320–30.

Hofweber, Thomas. 2005. Number determiners, numbers, and arithmetic. *Philosophical Review* 114: 179–225.

2007. Innocent statements and their metaphysically loaded counterparts. *The Philosophers Imprint* 7(1).

Hon, Giora and Goldstein, Bernard. 2008. *From* Summetria *to* Symmetry: *The Making of a Revolutionary Scientific Concept*. Berlin: Springer.

Horsten, Leon. 2007. Philosophy of mathematics. In *Stanford Encyclopedia of Philosophy*, ed. E. N. Zalta (winter 2007 edn). Available at: http://plato.stanford.edu/archives/win2007/entries/philosophy-mathematics/.

Høyrup, Jens. 1990. Sub-scientific mathematics: observations on a pre-modern phenomenon. *History of Science* 28: 63–86.

2005. Leonardo Fibonacci and the abbaco culture: a proposal to invert the roles. *Revue d'Histoire des Mathématiques* 11: 23–56.

n.d. For much more work, and difficult-to-locate or unpublished papers on *abbaco* culture, see http://akira.ruc.dk/~jensh/Selected%20themes/Abbacus %20mathematics/index.htm.

Husserl, Edmund. 1970. *The Crisis of European Sciences and Transcendental Phenomenology: An Introduction to Phenomenological Philosophy*, trans. from the German edition of 1954 by David Carr. Evanston, IL: Northwestern University Press.

Huizinga, Johan. 1949. *Homo Ludens: A Study of the Play-element in Culture*. London: Routledge. Trans. from German (Swiss) edition of 1940. Dutch original online in Digitale Biblioteek voor de Nederlandse Letteren, DBNL.

Institute for Advanced Study. 2010. The fundamental lemma: from minor irritant to central problem. *The Institute Letter*. Princeton, NJ: IAS, pp. 1, 4, 5, 7.

Jaffe, Arthur and Quinn, Frank. 1993. 'Theoretical mathematics': toward a cultural synthesis of mathematics and theoretical physics. *Bulletin of the American Mathematical Society* 29: 1–13.

Johnson, W. E. 1892. The logical calculus 1. General principles. *Mind* new series 1: 3–30.

Kahn, Charles. 2001. *Pythagoras and Pythagoreans*. Indianapolis: Hackett.

Kanovei, Vladimir, Katz, Mikhail G., and Mormann, Thomas. 2013. Tools, objects, and chimeras: Connes on the role of hyperreals in mathematics. *Foundations of Science* 18(2): 259–96.

Kant, Immanuel. 1929. *Critique of Pure Reason*, trans. Norman Kemp Smith. London: Macmillan.

1992. Concerning the ultimate ground of the differentiation of directions in space. In D. Walford and R. Meerbote, *Immanuel Kant: Theoretical Philosophy 1775–1770*, Cambridge University Press, 365–72. Original 1768.

1997a. *Prolegomena to Any Future Metaphysics that Will Be Able to Come Forward as Science*, trans. Gary Hatfield. Cambridge University Press.

1997b. *Lectures on Metaphysics*, trans. K. Ameriks and S. Naragon. Cambridge University Press.

1998. *Critique of Pure Reason*, trans. Paul Guyer and Allen Wood. Cambridge University Press.

Kaplan, David. 1975. How to Russell a Frege-Church. *Journal of Philosophy*, 772: 716–29.

Kapustin, Anton and Witten, Edward. 2006. Electric-magnetic duality and the geometric Langlands program. Available at: http://arXiv:hep-th/0604151.pdf.

Kästner, Abraham Gotthelf. 1780. *Anfangsgründe der angewandten Mathematik. 1. Mechanische und Optische Wissenschaften*. Göttingen: Vandenhoeck.

Kennedy, Christopher and Stanley, Jason. 2009. On 'average'. *Mind* 118: 583–646.

Kitcher, Philip. 1983. *The Nature of Mathematical Knowledge*. Oxford University Press.

Klein, Jacob. 1968. *Greek Mathematical Thought and the Origin of Algebra*, trans. E. Brann. Cambridge MA: MIT Press.

Kline, Morris. 1953. *Mathematics in Western Culture*. Oxford University Press.
　　1980. *Mathematics: The Loss of Certainty*. Oxford University Press.

Kochan, Jeff. 2006. Rescuing the Gorgias from Latour. *Philosophy of the Social Sciences* 36: 395–422.

Knorr, Wilbur. 1975. *The Evolution of the Euclidean Elements: A Study of the Theory of Incommensurable Magnitudes and its Significance for Early Greek Geometry*. Dordrecht: Reidel.
　　1986. *The Ancient Tradition of Geometric Problems*. Boston: Birkhäuser.

Kremer, Michael. 2008. Soames on Russell's logic: a reply. *Philosophical Studies* 139: 209–12.

Kreisel, Georg. 1958. Wittgenstein's remarks on the *Foundations of Mathematics*. *British Journal for the Philosophy of Science*, 9: 135–58.

Krieger, Martin H. 1987. The physicist's toolkit. *American Journal of Physics* 55: 1033–8.
　　1992. *Doing Physics: How Physicists Take Hold of the World*. Bloomington and Indianapolis: Indiana University Press.

Kripke, Saul. 1965. Semantical analysis of intuitionistic logic. In J. Crossley and M. A. E. Dummett (eds.) *Formal Systems and Recursive Functions*. Amsterdam: North-Holland, 92–130.
　　1980. *Naming and Necessity*. Oxford: Blackwell.

Kuhn, Thomas S. 1962. *The Structure of Scientific Revolutions*. University of Chicago Press.
　　1977. Second thoughts on paradigms. In Thomas S. Kuhn, *The Essential Tension: Selected Studies in Scientific Tradition and Change*. University of Chicago Press, 293–319; first published 1974.

Ladyman, James. 2009. Structural realism. In *Stanford Encyclopedia of Philosophy*, ed. Edward N. Zalta (summer 2009 edn). Available at: http://plato.stanford.edu/archives/sum2009/entries/structural-realism/.

Laird, W. R. 1997. Galileo and the mixed sciences. In Daniel A. Di Liscia *et al.* (eds.) *Method and Order in the Renaissance Philosophy of Nature*. Aldershot: Ashgate, 253–70.

Lakatos, Imre. 1963–4. Proofs and refutations. *British Journal for the Philosophy of Science*, 15: Part I, 1–25, Part II, 120–39, Part III, 221–45, Part IV, 296–342.
　　1976. *Proofs and Refutations: The Logic of Mathematical Discovery*. Expanded version of Lakatos (1963–4), ed. John Worrall and Elie Zahar. Cambridge University Press.

Langlands, Robert. 2010. Is there beauty in mathematical theories? Available at: http://publications.ias.edu/sites/default/files/ND.pdf.
　　n.d. Website collecting Langlands' work. Available at: http://sunsite.ubc.ca/DigitalMathArchive/Langlands/, but is now kept up at http://publications.ias.edu/rpl/.

Latour, Bruno. 1999. *Pandora's Hope: Essays on the Reality of Science Studies.* Cambridge, MA: Harvard University Press.

2008. The Netz-works of Greek deductions. *Social Studies of Science* 38: 441–59.

Lennox, James G. 1986. Aristotle, Galileo, and 'mixed sciences'. In W. Wallace (ed.) *Reinterpreting Galileo.* Washington, DC: Catholic University of America, 29–51.

Lévi-Strauss, Claude. 1954. The mathematics of man. *International Social Science Bulletin* 6: 581–90.

1969. *The Elementary Structures of Kinship.* London: Eyre and Spottiswoode. French original: *Les structures élémentaires de la parenté*, Paris: Presses Universitaires de France, 1949.

Liggins, David. 2008. Quine, Putnam, and the 'Quine–Putnam' indispensability argument. *Erkenntnis* 68: 113–27.

Littlewood, J. E. 1953. *A Mathematician's Miscellany.* London: Methuen.

Linsky, Bernard. 2009. Logical constructions. In *Stanford Encyclopedia of Philosophy*, ed. Edward N. Zalta (winter 2009 edn). Available at: http://plato.stanford.edu/archives/win2009/entries/logical-construction/.

Lloyd, Geoffrey. 1990. *Demystifying Mentalities.* Cambridge University Press.

Lucas, John. 1961. Minds, machines and Gödel. *Philosophy* 36: 112–27.

Lyotard, Jean-François. 1984. *The Postmodern Condition: A Report on Knowledge.* Minneapolis: University of Minnesota Press. Translated from the French of 1979.

MacFarlane, John. 2009. Logical constants. In *Stanford Encyclopedia of Philosophy*, ed. Edward N. Zalta (fall 2009 edn). Available at:http://plato.stanford.edu/archives/fall2009/entries/logical-constants/.

MacKenzie, Donald. 1993. Negotiating arithmetic, constructing proof: the sociology of mathematics and information technology. *Social Studies of Science* 23(1): 27–65.

2001. *Mechanizing Proof: Computing, Risk, and Trust.* Cambridge, MA: MIT Press.

Maddy, Penelope. 1997. *Naturalism in Mathematics.* Oxford University Press.

1998. How to be a naturalist about mathematics. In H. G. Dales and G. Oliveri (eds.) *Truth in Mathematics.* Oxford University Press.

2001. Naturalism: friends and foes. *Philosophical Perspectives* 15: 37–67.

2005. Three forms of naturalism. In S. Shapiro (ed.) *Oxford Handbook of the Philosophy of Mathematics and Logic.* Oxford University Press, 437–59.

2007. *Second Philosophy: A Naturalistic Method.* Oxford University Press.

2008. How applied mathematics became pure. *Review of Symbolic Logic* 1: 16–41.

Mancosu, Paolo. 1998. *From Brouwer to Hilbert: The Debate on the Foundations of Mathematics in the 1920s.* Oxford University Press.

2009a. Measuring the size of infinite collections of natural numbers: was Cantor's theory of infinite number inevitable? *Review of Symbolic Logic* 2: 612–46.

2009b. Mathematical style. In *Stanford Encyclopedia of Philosophy*, ed. Edward N. Zalta (spring 2010 edn). Available at: http://plato.stanford.edu/archives/spr2010/entries/mathematical-style/.

Manders, Kenneth. 1989. Domain extension and the philosophy of mathematics. *Journal of Philosophy* 86: 553–62.

Martin-Löf, Per. 1996. On the meanings of the logical constants and the justifications of the logical laws. *Nordic Journal of Philosophical Logic* 1: 11–60.

Mashaal, Maurice. 2006. *Bourbaki: A Secret Society of Mathematicians*. Providence, RI: American Mathematical Society. Translation of *Bourbaki: une société secrète de mathématiciens*. Paris: Belin, 2002.

Masters, Alexander. 2011. *Simon: The Genius in my Basement*. London: Fourth Estate.
   2012. Meet the mathematical 'genius in my basement'. An interview on 26 February with (US) National Public Radio. Available at: npr.org.

Mazur, Barry. n.d. Questions about number. Available at: http://www.math.harvard.edu/~mazur/papers/scanQuest.pdf.

Maxwell, James Clerk. 1882. Does the progress of Physical Science tend to give any advantage to the opinion of Necessity (or Determinism) over that of the Contingency of Events and the Freedom of the Will? (read 11 February 1873). In L. Campbell and W. Garnett, *The Life of James Clerk Maxwell, with Selections from his Correspondence and Occasional Writings*. London: Macmillan, 434–44.

Melia, Joseph. 1995. On what there's not. *Analysis* 55: 223–9.
   2000. Weaseling away the indispensability argument. *Mind* 109: 455–79.

Mews, Constant J. 1992. Nominalism and theology before Abelard: new light on Roscelin of Compiègne. *Vivarium* 30: 4–33.
   2002. *Reason and Belief in the Age of Roscelin and Abelard*. Aldershot: Ashgate.

Mialet, Hélène. 2012. *Hawking Incorporated: Stephen Hawking and the Anthropology of the Knowing Subject*. University of Chicago Press.

Mill, John Stuart. 1965–83. *Collected Works of John Stuart Mill*, ed. J. Robson, 28 vols. Toronto University Press.

Minkowski, Hermann. 1909. Raum und Zeit. *Jahresberichte der Deutschen Mathematiker-Vereinigung* 20: 1–14. Translated in *The Principle of Relativity: A Collection Of Original Memoirs on the Special and General Theory of Relativity*. London: Methuen, 1923.

Moore, G. E. 1900. Necessity. *Mind* new series 9: 283–304.

Mühlhölzer, Felix. 2006. 'A mathematical proof must be surveyable' – what Wittgenstein meant by this and what it implies. *Grazer Philosophische Studien* 71: 57–86.

Murakami, Haruki. 2011. *IQ84*. New York: Knopf. Translated from the Japanese original of 2009–10.

Ne'eman, Yuval. 2000. Pythagorean and Platonic conceptions of xxth century physics. In N. Alon *et al.* (eds.) *GAFA 2000: Visions in Mathematics Towards 2000*. Basel: Birkhäuser, 383–405.

Nelson, Edward. 1977. Internal set theory: a new approach to nonstandard analysis. *Bulletin of the American Mathematical Society* 83: 1165–98.

Netz, Reviel. 1999. *The Shaping of Deduction in Greek Mathematics: A Study in Cognitive History*. Cambridge University Press.
   2009. *Ludic Proof: Greek Mathematics and the Alexandrian Aesthetic*. Cambridge University Press.

Netz, Reviel and Noel, William. 2007. *The Archimedes Codex: Revealing the Secrets of the World's Greatest Palimpsest*. London: Wiedenfeld and Nicolson.

Núñez, Rafael. 2011. No innate number line in the human brain. *Journal of Cross-Cultural Psychology* 45: 651–68.

Parsons, Charles. 1964. Frege's theory of number. In Max Black (ed.) *Philosophy in America*. New York: Cornell University Press, 180–203.

1990. The structuralist view of mathematical objects. *Synthese* 84: 303–46.

2008. *Mathematical Thought and its Objects*. Cambridge University Press.

Peirce, Benjamin. 1881. Linear associative algebra. *American Journal of Mathematics* 4: 97–229.

Peirce, Charles Sanders. 1931–58. *Collected Papers*, ed. Charles Hartshorne and Paul Weiss, 8 vols. Cambridge, MA: Harvard University Press. References are given by volume and paragraph number, rather than pages, thus *CP* 4: 235 (1902) means paragraph 235 in vol. 4, dating from 1902.

Pierce, Roy. 1966. *Contemporary French Political Thought*. Oxford University Press.

Penrose, Roger. 1997. *The Large, the Small and the Human Mind*. Cambridge University Press.

Pickering, Andrew. 1995. *The Mangle of Practice: Time, Agency, and Science*. University of Chicago Press.

Plato. 1997. *Plato: Complete Works*, ed. John M. Cooper. Indianapolis: Hackett.

Polya, Georg. 1948. *How to Solve It: A New Aspect of Mathematical Method*. Princeton University Press. Reprinted 2004 with a valuable foreword by John Conway.

1954. *Mathematics and Plausible Reasoning*, vol. 1: *Induction and Analogy in Mathematics*. Princeton University Press.

Poncelet, Jean-Victor. 1822. *Traité des propriétés projectives des figures*. Paris: Bachelier.

1836. *Expériences faites à Metz, en 1834 … sur les batteries de brèche, sur la pénétration des projectiles dans divers milieux résistans, et sur la rupture des corps par le choc, suivies du rapport fait, sur ces expériences, à l'Académie des sciences … le 12 octobre 1835*. Paris: Corréard jeune.

1839. *Introduction à la mécanique industrielle, physique ou expérimentale*. Metz: Mme Thiel; Paris: Eugène André; first edn 1829.

Popper, Karl. 1962. *The Open Society and its Enemies*, vol. 1: Plato. London: Routledge; first edn 1945.

Putnam, Hilary. 1971. *Philosophy of Logic*. New York: Harper. Reprinted in Putnam (1979).

1975a. What is mathematical truth? In Hilary Putnam, *Philosophical Papers*, vol. 1: *Mathematics, Matter and Method*. Cambridge University Press, 60–78.

1975b. The meaning of 'meaning'. In Hilary Putnam, *Philosophical Papers*, vol. II: *Mind, Language and Reality*. Cambridge University Press, 215–71.

1979. Philosophy of logic. In Hilary Putnam, *Philosophical Papers*, vol. 1: *Mathematics, Matter and Method*, second edn. Cambridge University Press, 323–57. Reprint of Putnam (1971).

1982. Peirce the logician. *Historia Mathematica* 9: 290–301. Reprinted in his *Realism with a Human Face*. Cambridge, MA: Harvard University Press, 1990, 252–60.

1994. Rethinking mathematical necessity. In *Words and Life* ed. James Conant. Cambridge, MA: Harvard University Press, 245–63.

Quine, W. V. 1936. Truth by convention. In Otis H. Lee (ed.) *Philosophical Essays for A. N. Whitehead*. New York: Russell and Russell, 90–124. Original reprinted in Benacerraf and Putnam (1964: 322–45); reprinted with mild revisions in W. V. Quine, *Ways of Paradox*. New York: Random House, 1966.

1950. *Methods of Logic*. New York: Holt.

1953. On what there is. In *From a Logical Point of View*. Cambridge, MA: Harvard University Press, 1–19. (From page 149 of the book: this paper first appeared in the *Review of Metaphysics* in 1948, earlier versions having been presented as lectures at Princeton and Yale in March and May of that year. It lent its title to a symposium at the joint session of the Aristotelian Society and the Mind Association at Edinburgh, July 1951, along with the animadversions of the symposiasts, in the Aristotelian Society's supplementary volume for 1951.)

1960. *Word and Object*. Cambridge, MA: MIT Press.

1967. Accompanying note on Whitehead and Russell on Incomplete Symbols. In van Heijenoort (1967: 216–17).

1969. Epistemology naturalized. In *Ontological Relativity, and Other Essays*. New York: Columbia University Press, 69–90.

2008. Naturalism; or, living within one's means. In D. Føllesdal and D. B. Quine (eds.) *Confessions of a Confirmed Extensionalist and Other Essays*. Cambridge, MA: Harvard University Press, 461–72; reprinted from *Dialectica* 49 (1995): 251–61.

Ramsey, F. P. 1950. *The Foundations of Mathematics and other Logical Essays*. London: Routledge.

Reck, Erich. 2003. Dedekind's structuralism: an interpretation and partial defense. *Synthese* 137: 369–419.

2011. Dedekind's contributions to the foundations of mathematics. In *Stanford Encyclopedia of Philosophy*, ed. Edward N. Zalta (fall 2011 edn). Available at: http://plato.stanford.edu/archives/fall2011/entries/dedekind-foundations/.

Reck, Erich and Price, Michael. 2000. Structures and structuralism in contemporary philosophy of mathematics. *Synthese* 125: 341–83.

Resnik, Michael. 1981. Mathematics as a science of patterns: ontology and reference. *Noûs* 15: 529–50.

1988. Mathematics from the structural point of view. *Revue Internationale de Philosophie* 42: 400–24.

Robinson, Abraham. 1966. *Non-Standard Analysis*. Amsterdam: North-Holland.

Ronan, Mark. 2006. *Symmetry and the Monster: One of the Greatest Quests in Mathematics*. Oxford University Press.

Rosen, Joe. 1998. *Symmetry Discovered: Concepts and Applications in Nature and Science*. Minneola, NY: Dover; first edn 1975. Cambridge University Press.

Rothstein, Edward. 2010. 'Before Pythagoras' at New York University. *New York Times*, 26 November.

Russell, Bertrand. 1900. *A Critical Exposition of the Philosophy of Leibniz*. Cambridge University Press.

1903. *The Principles of Mathematics*. Cambridge University Press.

1905. On denoting. *Mind* new series 14: 479–93.

1912. *The Problems of Philosophy*. London: Williams and Norgate; The Home University Library of Modern Knowledge.

1917. The relation of sense data to physics. In Bertrand Russell, *Mysticism and Logic*. London: Allen and Unwin, 145–79.

1924. Logical atomism. In J. H. Muirhead, *Contemporary British Philosophy: Personal Statements*, first series. London: Allen and Unwin, 359–84.

1967. *The Autobiography of Bertrand Russell*, vol. 1. London: Allen and Unwin.

Scharlau, Walter. 2008. Who is Alexander Grothendieck? *Notices of the American Mathematical Society* 55: 930–41.

Schärlig, Alain. 2001. *Compter avec des cailloux: le calcul élémentaire sur l'abaque chez les anciens Grecs*. Lausanne: Presses Polytechniques et Universitaires Romandes.

Schmandt-Besserat, Denise. 1992. *Before Writing*, vol. 1: *From Counting to Cuneiform*. Austin: University of Texas Press.

Sebald, W. G. 1996. *The Emigrants*. New York: New Directions. (*Die Ausgewanderten*. Frankfurt am Main: Eichborn, 1993.)

Serre, Jean-Pierre. 2004. Interview with J.-P. Serre, by Martin Raussen and Christian Skau. *Notices of the American Mathematical Society* 51: 210–14.

Shapiro, Stewart. 1997. *Philosophy of Mathematics: Structure and Ontology*. Oxford University Press.

2000. *Thinking about Mathematics: The Philosophy of Mathematics*. Oxford University Press.

Shen, Kangsheng, Crossley, John N., and Lun, Anthony W.-C. 1999. *The Nine Chapters on the Mathematical Art: Companion and Commentary*. Oxford University Press.

SIAM (Society for Industrial and Applied Mathematics) n.d.. Website available at: www.siam.org/about/more/whatis.php.

Siegmund-Schultze, Reinhard. 2004. A non-conformist longing for unity in the fractures of modernity: towards a scientific biography of Richard von Mises. *Science in Context* 17: 333–70.

Soames, Scott. 2008. No class: Russell on contextual definition and the elimination of sets. *Philosophical Studies* 139: 213–18.

Sober, Elliott. 1993. Mathematics and indispensability. *Philosophical Review*, 102: 35–57.

Solomon, Ronald. 2005. Review of Aschbacher and Smith 2004. *Bulletin of the American Mathematical Society* new series 43: 115–21.

Spencer, Joel and Graham, Ronald. 2009. The elementary proof of the prime number theorem. *The Mathematical Intelligencer* 31: 18–23.

Sprague, Roland. 1939. Beispiel einer Zerlegung des Quadrats in lauter verschiedene Quadrate. *Mathematische Zeitschrift* 45: 607f.

Stein, Howard. 1988. Logos, logic, and Logistiké: some philosophical remarks on nineteenth-century transformation of mathematics. In Aspray and Kitcher (1988: 238–59).

Steiner, Mark. 1973. *Mathematical Knowledge*. New York: Cornell University Press.

1998. *The Applicability of Mathematics as a Philosophical Problem*. Cambridge, MA: Harvard University Press.

2009. Empirical regularities in Wittgenstein's philosophy of mathematics. *Philosophia Mathematica* 17: 1–34.

Stewart, Ian. 2007. *Why Beauty is Truth: A History of Symmetry*. New York: Basic Books.

Stigler, Stephen. 1980. Stigler's law of eponymy. *Transactions of the New York Academy of Sciences*, 39: 147–58.

Strawson, P. F. 1950. On referring. *Mind* 59: 320–44.

Tait, William. 1996. Frege versus Carnap and Dedekind: on the concept of number. In W. Tait (ed.), *Early Analytic Philosophy: Frege, Russell, Wittgenstein: Essays in Honor of Leonard Linsky*. Chicago: Open Court, 213–48.

2001. Beyond the axioms: the question of objectivity in mathematics. *Philosophia Mathematica* 9: 21–36.

2006. Review of Gödel's correspondence on proof theory and constructive mathematics in the Collected Works. *Philosophia Mathematica* 14: 76–111.

Tappenden, Jamie. 2009. Mathematical concepts and definitions. In Paolo Mancosu (ed.) *Philosophy of Mathematical Practice*. Oxford University Press, 256–75.

Tegmark, Max. 2008. The mathematical universe. *Foundations of Physics* 38: 101–50.

Thomson, William (Lord Kelvin) and Tait, Peter Guthrie. 1867. *A Treatise of Natural Philosophy*. Oxford: Clarendon Press.

Thurston, William P. 1994. On proof and progress in mathematics. *Bulletin of the American Mathematical Society* 30: 161–177. (Anthologized in Tymoczko (1998) and Hersh (2006).)

Toledo, Sue. 2011. Sue Toledo's notes of her conversations with Gödel in 1972–5. In Juliette Kennedy and Roman Kossak (eds.) *Set Theory, Arithmetic, and Foundations of Mathematics: Theorems, Philosophies*. Cambridge University Press, 200–7.

Tutte, William. 1958. Originally in Martin Gardner's 'Mathematical Games' column, *Scientific American* 199 (November): 136–44; reprinted as 'Squaring the square', in Martin Gardner, *More Mathematical Puzzles and Diversions*. London: Penguin, 1961.

Tymoczko, Thomas (ed.). 1998. *New Directions in the Philosophy of Mathematics. An Anthology*, revised and expanded edn. Princeton University Press.

Van Bendegem, J. P. 1998. What, if anything, is experiment in mathematics? In D. Anapolitanos *et al.* (eds.) *Philosophy and the Many Faces of Science*. London: Rowman and Littlefield.

Van Heijenoort, Jean. 1967. *From Frege to Gödel: A Source Book in Mathematical Logic, 1879–1931*. Cambridge, MA: Harvard University Press.

Voevodky, Vladimir. 2010a. Univalent foundations project. Available at: www.math.ias.edu/~vladimir/Site3/Univalent_Foundations_files/univalent_founda tions_project.pdf.

2010b. What if current foundations of mathematics are inconsistent? Available at: http://video.ias.edu/voevodsky-80th.

Waismann, Friedrich. 1945. Verifiability. *Proceedings of the Aristotelian Society*, supplementary vol. 19: 119–50.

Wang, Hao. 1961. Proving theorems by pattern recognition II. *Bell Systems Technical Journal* 40: 1–41.

Weber, Heinrich. 1893. Leopold Kronecker. *Mathematische Annalen* 43: 1–26. Reprinted from *Jahresbericht der deutschen Mathematiker-Vereinigung* 2 (1891–2): 5–31.

Weil, André. 2005. A 1940 letter of André Weil on analogy in mathematics, trans. Martin Krieger. *Notices of the American Mathematical Society* 52: 334–41. Original in Weil, *Œuvres scientifiques: Collected Papers*. New York: Springer, vol. 1, 244–55.

Weinberg, Steven. 1997. Changing attitudes and the standard model. In Lillian Hoddeson *et al.* (eds.) *The Rise of the Standard Model: Particle Physics in the 1960s and 1970s*. Cambridge University Press, 36–44.

Weyl, Hermann. 1921. Über die neue Grundlagenkrise der Mathematik. *Mathematische Zeitschrift* 10: 39–79. Translation entitled 'On the new foundational crisis in mathematics', in Mancosu (1998: 86–118).

1952. *Symmetry*. Princeton University Press.

1987. *The Continuum: A Critical Examination of the Foundation of Analysis*, trans. from the German of 1918. Kirksville, MO: Thomas Jefferson University Press.

2009. *Mind and Nature: Selected Writings on Philosophy, Mathematics and Physics*, ed. and intro. Peter Pesic. Princeton University Press.

Whately, Richard. 1831. *Elements of Logic Comprising the Substance of the Article in the Encyclopaedia Metropolitana*, fourth revised edn. London: B. Fellowes.

1874. *An Abstract of Bishop Whately's Logic, to the End of chapter 3, book 2, with Examination Papers*, ed. Bion Reynolds. Cambridge: W. Tomlin.

Whitehead, A. N. 1926. Mathematics as an element in the history of thought. Lecture 2 of the Lowell Lectures, 1925. In *Science and the Modern World*. Cambridge University Press.

Wigner, Eugene. 1960. The unreasonable effectiveness of mathematics in the natural sciences. *Communications in Pure and Applied Mathematics*, 13: 1–14.

Wilson, Mark. 1992. Frege: the royal road from geometry. *Noûs* 26: 149–80.

2000. The unreasonable uncooperativeness of mathematics in the natural sciences. *The Monist* 86: 296–314.

Wittgenstein, Ludwig. 1922. *Tractatus Logico-Philosophicus*, trans. C. K. Ogden. London: Routledge and Kegan Paul.

1956. *Remarks on the Foundations of Mathematics*, ed. G. H. von Wright, R. Rhees, and G. E. M. Anscombe, trans. G. E. M. Anscombe. Oxford: Blackwell.

1961. *Tractatus Logico-Philosophicus*, trans. D. F. Pears and B. F. McGuinness. London: Routledge and Kegan Paul.

1969. *On Certainty*. Oxford: Blackwell.

1974. *Philosophical Grammar*, ed. Rush Rhees, trans. Anthony Kenny. Oxford: Blackwell.

1976. *Wittgenstein's Lectures on the Foundations of Mathematics, Cambridge, 1939: from the notes of R. G. Bosanquet, Norman Malcolm, Rush Rhees, and Yorick Smythies*, ed. Cora Diamond. University of Chicago Press.

1978. *Remarks on the Foundations of Mathematics*. ed. G. H. von Wright, R. Rhees, and G. E. M. Anscombe, trans. G. E. M. Anscombe, third revised edn. Oxford: Blackwell.

1979. *Wittgenstein's Lectures, Cambridge 1932–1935*, ed. Alice Ambrose. Totowa, NJ: Rowman and Littlefield.

2001. *Philosophical Investigations*, trans. G. E. M. Anscombe, third revised edn. Oxford: Blackwell.

Worrall, John. 1989. Structural realism: the best of both worlds? *Dialectica* 43: 99–124.

Wright, Crispin. 1983. *Frege's Conception of Numbers as Objects*. Aberdeen University Press.

Wronski, J. Hoëné de. 1811. *Introduction à la philosophie des mathématiques et technie d'algorithmie*. Paris: Courcier.

Wylie, Alexander. 1853. *Jottings on the Sciences of the Chinese*. Shanghai: Northern Herald.

Yablo, Stephen. 1998. Does ontology rest upon a mistake? *Proceedings of the Aristotelian Society*, supplement 72: 229–61.

Zeilberger, Doron. n.d. Opinions. Available at: www.math.rutgers.edu/~zeilberg/Opinion110.html.

Index

Printed in the United States
By Bookmasters